7/06

APR 05

DATE DUE

OCT 2 4 '05			

Carp rs
La

GAYLORD · PRINTED IN U.S.A.

Audel™

Carpenters and Builders Layout, Foundation, and Framing

All New 7th Edition

Mark Richard Miller
Rex Miller

WILEY
Wiley Publishing, Inc.

Vice President and Executive Group Publisher: Richard Swadley
Vice President and Publisher: Joseph B. Wikert
Executive Editor: Carol A. Long
Editorial Manager: Kathryn A. Malm
Development Editor: Kevin Shafer
Production Editor: Angela Smith
Text Design & Composition: TechBooks

Contents

Acknowledgments

No book can be written without the aid of many people. It takes a great number of individuals to put together the information available about any particular technical field into a book. Many firms have contributed information, illustrations, and analysis to the book.

The authors would like to thank every person involved for his or her contributions. Following are some of the firms that supplied technical information and illustrations.

American Plywood Association

Bilco Company

Billy Penn Gutters

National Forest Products Association

Owens-Corning

Portland Cement Association

Scholtz Homes, Inc.

Shetter-Kit, Inc.

Stanley Tools

Truswal Systems Corp.

Waco Scaffolding and Equipment

About the Authors

Mark Richard Miller finished his BS in New York and moved on to Ball State University, where he earned a master's degree, then went to work in San Antonio. He taught high school and finished his doctorate in College Station, Texas. He took a position at Texas A&M University in Kingsville, Texas where he now teaches in the Industrial Technology Department as a Professor and Department Chairman. He has co-authored 11 books and contributed many articles to technical magazines. His hobbies include refinishing a 1970 Plymouth Super Bird and a 1971 Road-runner.

Rex Miller was a Professor of Industrial Technology at The State University of New York, College at Buffalo for more than 35 years. He has taught at the technical school, high school, and college level for more than 40 years. He is the author or co-author of more than 100 textbooks ranging from electronics through carpentry and sheet metal work. He has contributed more than 50 magazine articles over the years to technical publications. He is also the author of seven Civil War regimental histories.

Introduction

The Audel Carpenters and Builders Layout, Foundation, and Framing: All New Seventh Edition is the third of four volumes that cover the fundamental tools, methods, and materials used in carpentry, woodworking, and cabinetmaking.

This volume was written for anyone who wants (or needs) to understand how the layout of a project or building is done; how the foundation of a house is constructed; and how to frame a house. The problems encountered here can make or ruin a house or any project. Problems encountered by the carpenter, woodworker, cabinetmaker, or do-it-yourselfer often need attention by someone familiar with the requirements of the job well-done. Whether remodeling an existing home or building a new one, the rewards of doing a good job are great.

This book has been prepared for use as a practical guide in the selection, maintenance, installation, operation, and repair of wooden structures. Carpenters and woodworkers (as well as cabinetmakers and new homeowners) should find this book (with its clear descriptions, illustrations, and simplified explanations) a ready source of information for the many problems that they might encounter while building, maintaining, or repairing houses and furniture. Both professionals and do-it-yourselfers who want to gain knowledge of woodworking and house building will benefit from the theoretical and practical coverage of this book.

This is the third of a series of four books in the *Carpenters and Builders Library* that was designed to provide you with a solid reference set of materials that can be useful both at home and in the field. Other books in the series include the following:

- *Audel Carpenters and Builders Tools, Steel Square, and Joinery: All New Seventh Edition*
- *Audel Carpenters and Builders Math, Plans, and Specifications: All New Seventh Edition*
- *Audel Carpenters and Builders Millwork, Power Tools, and Painting: All New Seventh Edition*

No book can be completed without the aid of many people. The Acknowledgments mention some of those who contributed to making this the most current in design and technology available to the carpenter. We trust you will enjoy using the book as much as we did writing it.

Chapter 1

Locating a Building

The term *layout* means the process used to locate and fix the reference lines that define the position of the foundation and outside walls of a building.

Selection of Site

Staking out (sometimes called a *preliminary layout*) is important. The exact location of the building has to be properly selected. It may be wise to dig a number of small, deep holes at various points. The holes should extend to a depth a little below the bottom of the basement.

If the holes extend down to its level, the groundwater (which is sometimes present near the surface of the earth) will appear in the bottom of the holes. This water will nearly always stand at the same level in all the holes.

If possible, a house site should be located so that the bottom of the basement is above the level of the groundwater. This may mean locating the building at some elevated part of the lot or reducing the depth of excavation. The availability of storm and sanitary sewers (and their depth) should have been previously investigated. The distance of the building from the curb is usually stipulated in city building ordinances, but this, too, should be known.

Staking Out

After the approximate location has been selected, the next step is to lay out the building lines. The position of all corners of the building must be marked in some way so that when the excavation is begun, workers will know the exact boundaries of the basement walls (see Figure 1-1). There are a couple of methods of laying out building lines:

- With surveyor's instrument
- By method of diagonals

The Lines

Several lines must be located at some time during construction, and they should be carefully distinguished. They include the following:

- The line of excavation that is the outside line
- The face line of the basement wall inside the excavation line

**Figure I-I   One way of laying out is with a hundred-foot tape.
Metal tape is standard, but this new fiberglass one works well
and cleans easily.** *(Courtesy of Stanley Tools.)*

- In the case of masonry building, the ashlars line that indicates the outside of the brick or stone walls

In a wooden structure, only the two outside lines must be located, and often the line of the excavation is determined at the outset.

Laying Out with Transit Instruments

A *transit* is an instrument of precision, and the work of laying out with this instrument is more accurate than with other methods. In Figure 1-2, let *ABCD* be a building already erected. At a distance from this (at right angle), building *GHJK* will be erected. Level up the instrument at point *E*, making *A* and *E* the distance the new building will be from points *A* and *B*. Make points *B* and *F* the same length as points *A* and *E*. At this point, drive a stake in the ground at point *G*, making points *F* and *G* the required distance between the two buildings. Point *H* will be on the same line as point *G*, making the distance between the two points as required.

Place the transit over point *G*, and level it up. Focus the transit telescope on point *E* or *F* and lock into position. Turn the horizontal

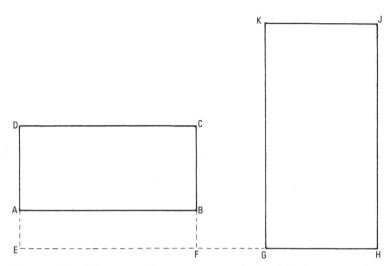

Figure 1-2   Diagram illustrating method of laying out with transit.

circle on the transit until one of the zeros exactly coincides with the vernier zero. Loosen the clamp screw and turn the telescope and vernier 90 degrees. This will locate point K, which will be at the desired distance from point G. For detailed operation of the transit (see Figure 1-3), see the manufacturer's instructions or information in the *Audel Carpenters and Builders Math, Plans, and Specifications* book of this series. (See the Introduction for more details on this series.) The level may be used in setting floor timbers, in aligning posts, and in locating drains.

Method of Diagonals
All that is needed in this method are a line, stakes, and a steel tape measure. Here, the right angle between the lines at the corners of a rectangular building is found by calculating the length of the diagonal that forms the hypotenuse of a right-angle triangle. By applying the following rule, the length of the diagonal (hypotenuse) is found.

Rule: The length of the hypotenuse of a right-angle triangle is equal to the square root of the sum of the squares of each leg.

Thus, in a right-angle triangle ABC, the hypotenuse is AC,

$$AC = \sqrt{AB^2 + BC^2}$$

Suppose, in Figure 1-4, *ABCD* represents the sides of a building to be constructed, and it is required to lay out these lines to the

Figure 1-3 Transit, used by builders, contractors, and others for setting grades, batter boards, and various earth excavations.

dimensions given. Substitute the values given in the previous equation as follows:

$$AC = \sqrt{30^2 + 40^2} = \sqrt{900 + 1600} = \sqrt{2500} = 50$$

To lay out the rectangle of Figure 1-4, first locate the 40-foot line AB with stake pins. Attach the line for the second side to B, and measure off this line the distance BC (30 feet), point C being indicated by a knot. This distance must be accurately measured with the line at the same tension as in A and B.

With the end of a steel tape fastened to stake pin A, adjust the position of the tape and line BC until the 50-foot division on the tape coincides with point C on the line. ABC will then be a right angle, and point C will be properly located.

The lines for the other two sides of the rectangle are laid out in a similar manner. After getting the positions for the corner stake pins, erect batter boards and permanent lines (see Figure 1-5). A simple procedure may be used in laying out the foundations for a small rectangular building. Be sure that the opposite sides are equal

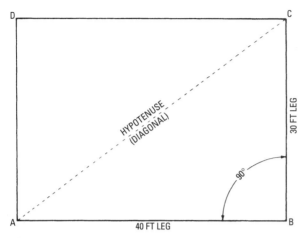

D ――――――――――――――――――――――――――― C

HYPOTENUSE (DIAGONAL)

30 FT LEG

90°

A ――――――――――――――――――――――――――― B

40 FT LEG

Figure 1-4 Diagram illustrating how to find the length of the diagonal in laying out lines of a rectangular building by using the diagonals method.

and then measure *both* diagonals. No matter what this distance may be, they will be equal if the building is square. No calculations are necessary, and the method is precise.

Points on Layout

For ordinary residence work, a surveyor or the city engineer is employed to locate the lot lines. Once these lines are established, the builder is able to locate the building lines by measurement.

A properly prepared set of plans will show both the contour of the ground on which the building is to be erected and the new grade line after the building is done. A convenient way of determining old grade lines and establishing new ones is by means of a transit, or with a Y level and a rod. Both instruments work on the same principle in grade work. As a rule, masonry contractors have their own Y levels and use them freely as walls are constructed, especially where levels are to be maintained as the courses of material are placed.

In locating the grade of the earth around a building, stakes are driven into the ground at frequent intervals and the amount of fill indicated by the heights of these stakes. Grade levels are usually established after the builders have finished, except that the mason will have the grade indicated where the wall above the grade is to be finished differently from the wall below grade. When a Y level is not

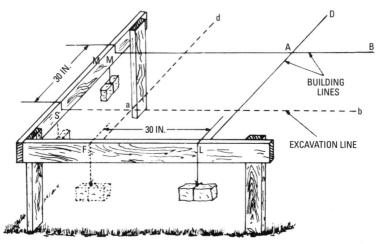

Figure 1-5 Permanent location of layout lines made by cutting in batter boards (boards marked S, M, F, L). Slits L and M locate the building lines. Approximately 30 inches away are lines F and S, which are excavation lines.

available, a 12- or 14-foot straightedge and a common carpenter's level may be used, with stakes being driven to "lengthen" the level.

Summary

The term *layout* means the process of locating a fixed reference line that will indicate the position of the foundation and walls of a building.

A problem sometimes encountered is groundwater. It is sometimes present near the surface of the earth and will appear in the bottoms of test holes, generally at the same level. If possible, a house should be located so that the bottom of the basement floor is above the level of the ground water.

After the location of the house has been selected, the next step is to lay out or stake out the building lines. The position of all corners of the house must be marked so that workers will know the exact boundaries of the basement walls.

There are several ways to lay out a building site. Two of these are with a surveyor's instrument and with diagonal measurements. When laying out a site, several lines must be located at some time during construction. These lines are the line of excavation (which is the outside line), the face line of the basement wall inside the excavation line, and, in the case of a masonry building, the ashlars line (which indicates the outside of the brick or stone wall).

Review Questions

1. What is groundwater?
2. Name two methods used in laying out a building site.
3. What is the difference between laying out and staking out?
4. What is the line of excavation?
5. What is the ashlars line?
6. What is the advantage of using a fiberglass measuring tape in the field?
7. How is a transit used in the layout of a basement?
8. What has to be done by the surveyor before the developer can lay out houses?
9. When are grade levels established?
10. What are batter boards?

Chapter 2

House Foundations

The *foundation* is the part of a building that supports the load of the superstructure. As generally understood, the term includes all walls, piers, columns, pilasters, and other supports below the first-floor framing.

Following are three general forms of foundation:

- Spread foundations (see Figure 2-1)
- Pile foundations
- Wood foundations

Spread foundations are the most popular type used. They receive the weight of the superstructure and distribute the weight to a stable soil base by means of individual footings. *Pile foundations*, on the other hand, transmit the weight of the superstructure through a weak soil to a more-stable base. Of relatively recent vintage is the all-weather all-wood foundation, which is made of plywood soaked with preservatives.

Following are three basic types of spread foundations:

- Slab-on-grade
- Crawl space
- Basement or full (see Figure 2-2)

Each foundation system is popular in certain geographic areas. The slab-on-grade is popular in the South and Southwest. The crawl space is popular throughout the nation. The basement is the most popular in the Northern states.

Slab-on-Grade

There are three basic types of *slab-on-grade*. The most popular is where the footing and slab are combined to form one integral unit. Another type has the slab supported by the foundation wall, and there is a type where the slab is independent of the footing and foundation wall (see Figure 2-3).

The procedure for constructing a slab-on-grade would be as follows:

1. *Clear the site*—In most cases, no excavation is needed, but some fill dirt may be needed. A tractor or bulldozer is usually used to remove the unnecessary brush and trees. It can also be used to spread the necessary fill.

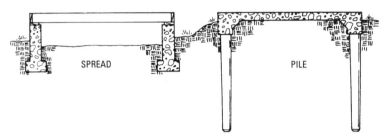

Figure 2-1   General forms for foundations.

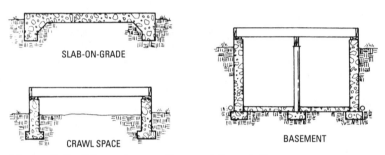

Figure 2-2   Three types of spread foundations.

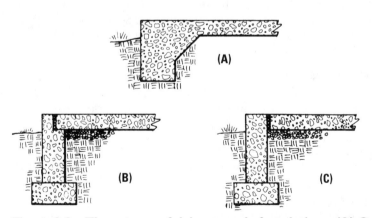

Figure 2-3   Three types of slab-on-grade foundations: (A) One integral unit, (B) supported by foundation wall; and (C) independent.

2. *Lay out the foundation*—This is usually done with batter board and strings. When the batter boards are attached to the stakes, the lowest batter board should be 8 inches above grade.

3. *Place and brace the form boards*—The forms are usually 2 × 12 boards, 2 × 6 boards, or 2 × 4 boards, and are aligned with the string. To keep the forms in proper position, they are braced with 2 × 4 boards. One 2 × 4 is placed adjacent to the form board and another is driven at an angle 3 feet from the form board. A "kicker" is placed between the two 2 × 4 boards, to tie the two together. These braces are placed around the perimeter of the building, 4 feet on center (see Figure 2-4).

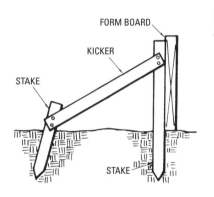

FORM BOARD

KICKER

STAKE

STAKE

Figure 2-4 Form construction.

4. *Additional fill is brought in*—The fill should be free of debris or organic matter and should be screeded to within 8 inches of the top of the forms. The fill should then be well tamped.

5. *Footings are dug*—The footings should be a minimum of 12 inches wide and should extend 6 inches into undisturbed soil. The footings should also extend at least one foot below the frost line (see Figure 2-5).

6. *Place the base course*—The base course is usually wash gravel or crushed stone and is placed 4 inches thick. The base course acts as a capillary stop for any moisture that might rise through the soil.

7. *Place the vapor barrier*—The vapor barrier is a sheet of 0.006 polyethylene and acts as a secondary barrier against moisture penetration.

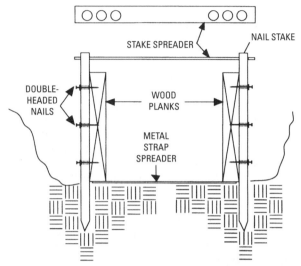

Figure 2-5   Nail stake footing forms are faster than all-wood forms, but require special equipment. High-carbon steel stakes are driven into the ground. Wood planks are nailed to the stakes. An adjustable metal stake spreader holds the top of the stakes together.

8. *Reinforce the slab*—In most cases, the slab is reinforced with 6 × 6 No. 10-gauge wire mesh. To ensure that the wire mesh is properly embedded, it is propped up or pulled up during the concrete pour. Fiberglass strands added to the concrete mix sometimes eliminate the need for wire mesh.

9. *Reinforce the footings*—The footings can be reinforced with three or four deformed metal bars 18 to 20 feet in length. The rods should not terminate at a corner. They should be bent to project around it. At an intersection of two rods, there should be an 18-inch overlap.

Once the forms are set and the slab bed completed, concrete is brought in and placed in position. The concrete should be placed in small piles and as near to its final location as possible. Small areas of concrete should be worked. (In working large areas; the water will supersede the concrete, causing inferior concrete.) Once the concrete has been placed in the forms, it should be worked (poked

Figure 2-6 Hand tamping, or jitterbugging, concrete to place large aggregate below the surface. The other worker (bent over) is screeding with a long 2 × 4 to proper grade.

and tamped) around the reinforcing bars and into the corners of the forms. If the concrete is not properly worked, air pockets or honeycombs may appear.

After the concrete has been placed, it must be struck or screeded to the proper grade. A long straightedge is usually used in the process. It is moved back and forth in a saw-like motion until the concrete is level with the forms. To place the large aggregate below the surface, the concrete is hand tamped, or *jitterbugged* (see Figure 2-6). A darby (a long flat tool for smoothing) is used immediately after the jitterbug and is also used to embed the large aggregate (see Figure 2-7). To produce a round on the edge of the concrete slab, an edger is used. The round keeps the concrete from chipping off and it increases the aesthetic appeal of the slab (see Figure 2-8). After the water sheen has left the surface of the slab, it is floated. Floating is used to remove imperfections and to compact the surface of the concrete. For a smooth and dense surface, the concrete is then troweled. It can be troweled with a steel hand trowel, or it can be troweled with a power trowel (see Figure 2-9).

Figure 2-7   The darby being used after the jitterbug process.

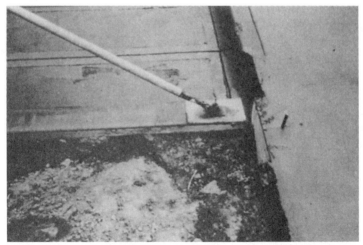

Figure 2-8   Using an edger to round off the edges.

Figure 2-9   Using a power trowel.

Once the concrete has been finished, it should be cured. There are three ways that the slab might be cured:

- Burlap coverings
- Sprinkling
- Ponding

Regardless of the technique used, the slab should be kept moist at all times.

The photographs shown in Figures 2-10 through 2-17 were taken where the soil permitted a shallow trench to serve as a footing for the slab-on-grade.

Crawl Space

A *crawl space* foundation system can be constructed of an independent footing and concrete-block foundation wall, or the footing and foundation wall can be constructed of concrete.

The footing should be constructed of concrete and should be placed below the frost line. The projection of the footing past the foundation wall should equal one-half the thickness of the

Figure 2-10 **Batter boards with string marking outer limits of the slab.**

Figure 2-11 **Forms started using string as a guide.**

Figure 2-12   Trenches for footings.

Figure 2-13   Tractor with trencher attached.

Figure 2-14   The forms in place. Plumbing vents and drains set in place.

Figure 2-15   Cables in place for the slab. Note trenching for plumbing and for footings and reinforcement for the slab. Site ready for concrete pouring.

Figure 2-16   Pouring the slab and leveling the concrete in the forms.

Figure 2-17   Concrete slab poured. Rebar for porch ready for concrete.

foundation wall (see Figure 2-18). The thickness of the footing should equal the width of the foundation wall. There are two basic ways to form a footing for a crawl space:

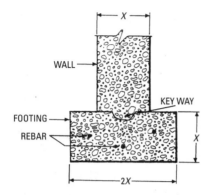

WALL

KEY WAY

FOOTING

REBAR

Figure 2-18 Footing construction. Keyways secure the foundation wall to the footing. This keeps the wall from sliding sideways from the pressure of backfill and helps slow down water seepage.

- Dig a footing trench and place the concrete in the trench. To maintain the proper elevation, grade stakes are placed in the trench.
- Use form boards. If form boards are used, they should be properly erected and braced. In some cases, additional strength may be needed, and reinforcement added.

The most convenient way to obtain concrete for a crawl space foundation or footing (or, for that matter, any job where a fair amount of material is required) is to have it delivered by truck. The mix will be perfect, and it will be poured exactly where you are ready for it with a minimum of effort on your part. Of course, this method is more expensive than you mixing it. If you mix it yourself, you can rent mixing equipment (such as a power mixer). A good, strong concrete mix is three bags of sand to every bag of cement, and enough water to keep the mix workable. On the other hand, you can use four bags of concrete sand (that is, sand with rocks in it) to every bag of cement. Forms are removed after the concrete has hardened. Before laying the concrete masonry, the top of the footings should be swept clean of dirt or loose material.

Regardless of whether the foundation wall is constructed of placed concrete or concrete blocks, the top should be a minimum of 18 inches above grade. This allows for proper ventilation, repair work, and visual inspection.

Basement Construction

In *basement construction*, foundation walls should be built with the utmost care and craftsmanship, because they are under great pressure from water in the ground (see Figure 2-19).

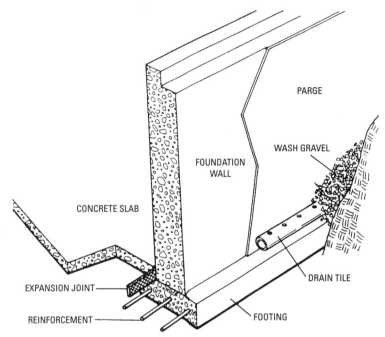

PARGE

WASH GRAVEL

FOUNDATION
WALL

CONCRETE SLAB

DRAIN TILE

EXPANSION JOINT

REINFORCEMENT

FOOTING

Figure 2-19   Basement construction.

To properly damp-proof the basement (if such a situation exists), a 4-inch drain tile can be placed at the base of the foundation wall. The drain tile can be laid with open joints or it can have small openings along the top. The tile should be placed in a bed of wash gravel or crushed stone and should drain into a dry streambed or storm sewer. The outside of the wall should then be parged, or covered with a mixture such as masonry cement, mopped with hot asphalt, or covered with polyethylene. These techniques will keep moisture from seeping through the foundation. For further protection, all surface water should be directed away from the foundation system. This can be done by ensuring that the downspout routes water away from the wall and that the ground slopes away from it.

Pile Foundation

Pile foundations are used to minimize and reduce settlement. There are two classifications of piles (see Figure 2-20):

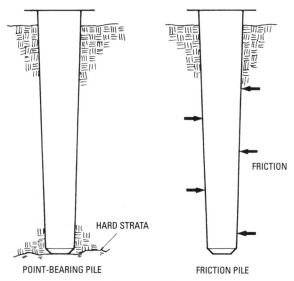

POINT-BEARING PILE   FRICTION PILE

Figure 2-20   Pile construction.

- *Point-bearing piles* transmit loads through weak soil to an area that has a better bearing surface.
- *Friction piles* depend on the friction between the soil and the pile to support the imposed load.

Many different kinds of materials are used for piles, but the most common are concrete, timber, and steel.

All-Weather Wood Foundation

The *wood foundation* is composed of wood and plywood soaked with preservatives. It was primarily developed so that a foundation could be installed in cold weather, when concrete cannot. The wood foundation is not difficult to install, and it is faster to install than a masonry foundation (see Figure 2-21). It can be used where working with concrete is limited by short building seasons. Wood foundations can be erected during freezing weather, or where there is too short a period to construct a different type of foundation.

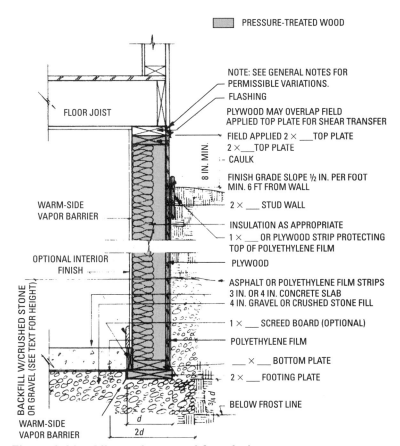

	PRESSURE-TREATED WOOD

FLOOR JOIST

NOTE: SEE GENERAL NOTES FOR
PERMISSIBLE VARIATIONS.
FLASHING
PLYWOOD MAY OVERLAP FIELD
APPLIED TOP PLATE FOR SHEAR TRANSFER
FIELD APPLIED 2 × ___TOP PLATE
2 ×___ TOP PLATE
CAULK
FINISH GRADE SLOPE ½ IN. PER FOOT
MIN. 6 FT FROM WALL
2 × ___ STUD WALL
INSULATION AS APPROPRIATE
1 × ___ OR PLYWOOD STRIP PROTECTING
TOP OF POLYETHYLENE FILM
PLYWOOD
ASPHALT OR POLYETHYLENE FILM STRIPS
3 IN. OR 4 IN. CONCRETE SLAB
4 IN. GRAVEL OR CRUSHED STONE FILL
1 × ___ SCREED BOARD (OPTIONAL)
POLYETHYLENE FILM
___ × ___ BOTTOM PLATE
2 × ___ FOOTING PLATE
BELOW FROST LINE

8 IN. MIN.

WARM-SIDE
VAPOR BARRIER

OPTIONAL INTERIOR
FINISH

BACKFILL W/CRUSHED STONE
OR GRAVEL (SEE TEXT FOR HEIGHT)

WARM-SIDE
VAPOR BARRIER

d
2d

Figure 2-21 All-weather wood foundation (*Courtesy of National Forest Products Assn.*)

Summary

There are three general forms of foundations: spread foundations, pile foundations, and wood foundations. Spread foundations are the most popular type used. There are three basic types of these: the slab-on-grade, the crawl space, and the basement.

The procedure for constructing a slab-on-grade would be to clear the site, lay out the foundation, place and brace the form boards, add fill, dig the footings, place the base course, place the vapor barrier, reinforce the slab, and reinforce the footings.

In basement construction, the foundation wall should be built with the utmost care and craftsmanship to resist the assault of ground water.

Review Questions

1. What are the two general forms of foundations?
2. What is a foundation?
3. Why is base course used?
4. How is a concrete slab cured?
5. How is a basement wall damp-proofed?
6. What does the word *parged* mean when working with concrete?
7. What is a darby?
8. Why are crawl spaces needed?
9. What are the two types of pile foundations?
10. Where would you use a wood foundation type of construction?

Chapter 3

Concrete Forms and Hardware

Since a concrete mixture is semi-fluid, it will take the shape of anything into which it is poured. Accordingly, molds or forms are necessary to hold the concrete to the required shape until it hardens. Such forms or molds may be made of metal, lumber, or plywood.

Formwork may represent as much as one-third of the total cost of a concrete structure, so the importance of the design and construction of this phase of a project cannot be overemphasized. The character of the structure, availability of equipment and form materials, anticipated repeat use of the forms, and familiarity with methods of construction influence design and planning of the formwork. Forms must be designed with knowledge of the strength of the materials and the loads to be carried. The ultimate shape, dimensions, and surface finish must also be considered in the preliminary planning phase.

Need for Strength

Forms for concrete structures must be tight, rigid, and strong. If forms are not tight, there will be a loss of concrete that may result in honeycombing, or a loss of water that causes sand streaking. The forms must be braced well enough to stay in alignment, and strong enough to hold the concrete. Keep in mind that concrete is heavy. Though structural concrete can vary in weight from 60 to 300 pounds per cubic foot (lb/ft^3), most structural slabs will use concrete weighing about 150 lb/ft^3. This includes the weight of the reinforcing. The form weights can vary from 4 to 15 pounds per square foot.

Special care should be taken in bracing and tying down forms such as those for retaining walls, in which the mass of concrete is large at the bottom and tapers toward the top. In this type of construction, and in other types (such as the first pour for walls and columns), the concrete tends to lift the form above its proper elevation. If the forms are to be used again, they must be designed so that they can easily be removed and re-erected without damage. Most forms are made of wood, but steel forms are commonly used for work involving large, unbroken surfaces such as retaining walls, tunnels, pavements, curbs, and sidewalks (see Figure 3-1). Steel forms for sidewalks, curbs, and pavements are especially advantageous, since they can be used many times.

Any concrete laid below ground level for support purposes (such as foundations) must start below the freeze line. This will vary for

Figure 3-1   Steel forms in place for a concrete slab.

different parts of the country, but is generally about 18 inches below ground level. The length of time necessary to leave the forms in place depends on the nature of the structure. For small-construction work where the concrete bears external weight, the forms may be removed as soon as the concrete will bear its own weight (that is, between 12 and 48 hours after the concrete has been poured). Where the concrete must resist the pressure of the earth or water (as in retaining walls or dams), the forms should be left in place until the concrete has developed nearly its final strength. This may be as long as three or four weeks if the weather is cold, or if anything else prevents quick curing.

Bracing

The bracing of concrete formwork falls into a number of categories. The braces that hold wall and column forms in position are usually 1-inch-thick boards or strips. Ordinarily such braces are not heavily stressed because the lateral pressure of the concrete is contained by wall ties and column clamps.

When it is not practical to use wall ties, the braces may be stressed depending on the height of the wall being poured. Braces of this kind are proportioned to support the wall forms against lateral forces. Deep beam and girder forms often require external braces to prevent the side forms from spreading.

The lateral bracing that supports slab forms is also important. Not only is it important in terms of safety, but also to prevent the distortions that can occur when the shores are knocked out of position. Lateral bracing should be left in place until the concrete is strong enough to support itself.

Economy

One concern of the builder, particularly today when costs of materials are so high, is economy. Forms for a concrete building, for example, can cost more than either the concrete or reinforcing steel, or, in some cases, more than both together. Hence, it is important to seek out every possible cost-cutting move.

Saving on costs starts with the design of the building. The designer must keep in mind the forms required for the building's construction and look to build in every possible economy. For example, it may be possible to adjust the size of beams and columns so that they can be formed with a combination of standard lumber or plywood-panel sizes.

For example, in making a beam 10 or 12 inches wide, specially ripped form boards are needed. On the other hand, surfaced 1-inch × 6-inch ($5^1/_2$-inch actual size) form boards will be just right for an 11-inch-wide beam. Experience has shown that it is entirely practical to design structures around relatively standardized beam and column sizes for the sake of economy. When such procedures are followed, any diminution of strength can be compensated for by increasing the amount of reinforcing steel used—and money will still be saved.

Another costly area to avoid is excessive design. It may just require a relatively small amount of concrete to add a certain look to a structure, but the addition can be very expensive in terms of overall cost.

Another economy may be found in the area of lumber lengths. Long lengths can often be used without trimming. Studs need not be cut off at the top or at a wall form, but can be used in random lengths to avoid waste. Random-length wales can also be used. Where a wall form is built in place, it does no harm if some boards extend beyond the length of the form.

Paradoxically, some very good finish carpenters are not good at form building because they spend a lot of time building the forms—neatness and exact lengths or widths are not required. Because they have overdone the form building, when it comes time to strip the form, there may be many nails to remove, and the job may be much more complicated.

Fastening and Hardware

All sorts of devices are available to simplify the building and stripping of forms. The simplest is the double-headed nail (see Figure 3-2). The chief advantage of these nails is that they can be pulled out easily because they are driven in only to the first head (see Figure 3-3). There are also many different varieties of column clamps, adjustable shores, and screw jacks. Instead of wedges, screw jacks are especially suitable when solid shores are used.

Figure 3-2   Double-headed scaffold, or framing, nail.

A number of wedges are usually required in form building (see Figure 3-4). Wedges are used to hold form panels in place, as well as to draw parts of forms into line, to adjust shores and braces. Usually the carpenter makes the wedges on the job. A simple jig can be rigged up on a table or a radial arm saw for cutting the wedges.

Form ties (available in many styles) are devices that support both sides of wall forms against the lateral pressure of concrete. Used properly, form ties practically eliminate external bracing and greatly simplify the erection of wall forms.

A simple tie is merely a wire that extends through the form, the ends of the wire double around a stud or wale on each side. Although low in cost, simple wire ties are not entirely satisfactory because, under pressure from the concrete, they cut into the wood members and cause irregularities in the wall. The most satisfactory ties can be partially or completely removed from the concrete after it has set and hardened completely.

Lumber for Forms

Some concrete building construction is done by using wooden or plywood forms. If kiln-dried lumber is used, it should be thoroughly wet before concrete is placed. This is important because the lumber will absorb water from the concrete, and if the forms are made tight (as they should be), the swelling from absorption can cause the forms to buckle or warp. Oiling or using special compounds on the inside of forms (as detailed later in this chapter) before use is recommended. This is especially true if the forms are to be used repeatedly. It prevents absorption of water, assists in keeping the forms in shape when not in use, and makes their removal from around the concrete much easier.

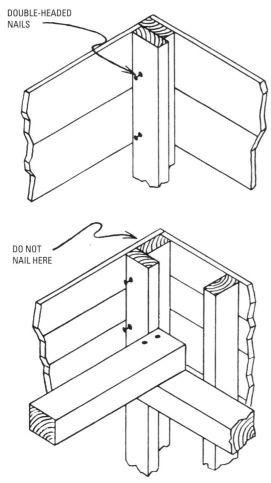

DOUBLE-HEADED
NAILS

DO NOT
NAIL HERE

Figure 3-3 Double-headed nails in action. Internal corners make stripping easier.

Spruce or pine seems to be the best all-around material. One or both woods can be obtained in most locations. Hemlock is not usually desirable for concrete formwork because it is liable to warp when exposed to concrete. One-inch lumber is normally sufficient for a building form, and so is $5/8$-inch, $1/2$-inch, and $3/4$-inch plywood.

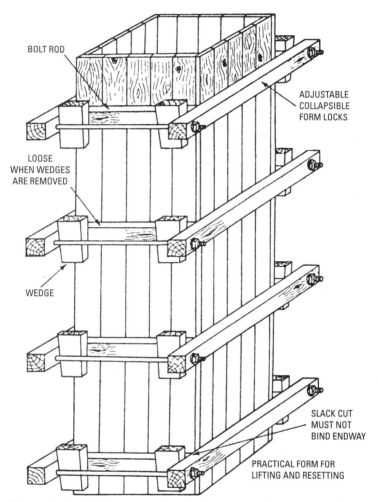

BOLT ROD

ADJUSTABLE
COLLAPSIBLE
FORM LOCKS

LOOSE
WHEN WEDGES
ARE REMOVED

WEDGE

SLACK CUT
MUST NOT
BIND ENDWAY

PRACTICAL FORM FOR
LIFTING AND RESETTING

Figure 3-4   Wedges are very important in form building.

It should be noted that the actual building of forms is often a complicated procedure, particularly when the form will be used in building a large structure (such as tall walls or large floors). Such things as the vertical pressure on the form when it is filled with concrete, the lateral pressure, and the amount of deflection of the form must all be considered. In essence, the form is a structure that

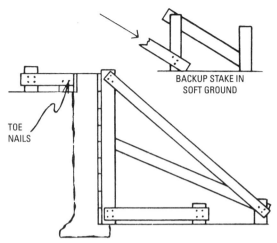

Figure 3-5 Cellar wall form in firm ground.

takes a lot of stress, and, therefore, it must be built to withstand that stress (see Figure 3-5).

Carpenters and builders must be familiar with the various methods and materials for building forms.

Practicality

If the job to be done is one on which speed is a prime factor (perhaps because of labor costs), then it is important that the forms not be unduly complicated or time-consuming. The design may be perfectly acceptable from an engineering standpoint but poor in terms of practicality.

One overall rule is that the components of the form be as large as can be handled practically. When possible, use panels. Panels may be made of suitably strong plywood, composed of boards all of the same width, and fastened together at the back with cleats.

In making forms for columns, the forms may be built as open-ended boxes. Of course, the panels cannot be completely nailed together until they are in place surrounding the column they are reinforcing. It is usually considered good practice to leave a clean-out hold on one side of the bottom of column forms (see Figure 3-6). Dirt, chips, and other debris are washed out of this hole, and then it is closed.

Even when column forms are prepared carefully, though, minor inaccuracies can occur. The form panels may swell, or may move

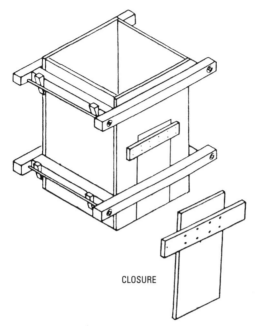

CLOSURE

Figure 3-6   Column form with clean-out.

slightly when the concrete is poured. In a well-designed and well-built form, such inaccuracies do not matter much because of the allowances built into the forms. As mentioned earlier, forms are wetted down thoroughly, thereby swelling them before use. It is doubly important to do this because of the water-cement ratio. If forms are dry, they will suck water from the concrete and change its properties. A form with a wet surface also solves the problem of wet concrete possibly honeycombing it.

When moldings are desired, they should be recessed into the concrete (see Figure 3-7). It is very simple to nail triangular strips or other molded shapes on the inside of a form (see Figure 3-8). Projection moldings on concrete require a recess in the form that is more difficult to make.

Horizontal recessed grooves should be beveled outward at the top and bottom, or at least at the bottom. A form for a flat ledge cannot readily be filled out with poured concrete. Additionally, ledges collect dirt and are likely to become unsightly.

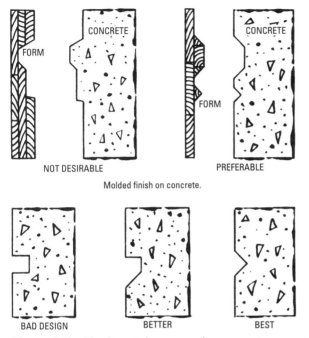

NOT DESIRABLE   PREFERABLE

Molded finish on concrete.

BAD DESIGN   BETTER   BEST

Figure 3-7 Horizontal grooves in concrete.

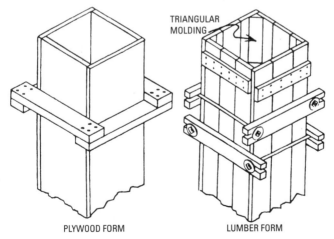

TRIANGULAR MOLDING

PLYWOOD FORM   LUMBER FORM

Figure 3-8 Column-form clamps.

If the placing of concrete can be timed to stop at a horizontal groove, that groove will serve to conceal the horizontal construction joint. One point (often overlooked) is the use of a horizontal construction joint as a ledge to support the forms for the next pour. This is a simple way to hold forms in alignment.

Texture

With forms, a variety of ways may be used to enhance the texture of the finished concrete. If a smooth finish is desired, forms may be lined with hardboard, linoleum, or similar materials. Small-headed nails should be used, and hammer marks should be avoided.

When a rough texture is wanted, rough form boards may be arranged vertically and horizontally, in alternate panels. Other arrangements (such as diamonds, herringbones, or chevrons) can also be used. As with almost anything else, the ingenious designer can come up with many ways to get the job done.

Size and Spacing

The size and spacing of ties are governed by the pressure of the concrete transmitted through the studs and wales to the ties. In other words, a wale acts as a beam, and the ties as reactions. It might be assumed, therefore, that large wales and correspondingly large ties should be used. However, if size and spacing are limited by economics, the tie-spacing limit is generally considered 36 inches, with 27 to 30 inches preferred.

Many tie styles make it necessary to place spreaders in the forms to keep the two sides of the form from being drawn together. Generally, spreaders should be removed so that they are not buried in the concrete. Traditional spreaders are quite good and are readily removed. Figure 3-9 shows how wood spreaders are placed and how they are removed.

Several styles of spreaders can be combined with ties. Some of these are made of wire nicked or weakened in such a way that the tie may be broken off in the concrete within an inch or two from the face of the form. After you withdraw the ties, the small holes that remain are easily patched with mortar.

Stripping Forms

Unlike most structures, concrete forms are temporary. The forms must later be removed or stripped (disassembled). Sometimes (for example, in building a single home) forms are used only once and then discarded. Nevertheless, in most cases, economy dictates that a form be used and reused. Indeed, economical heavy construction depends on reusing forms.

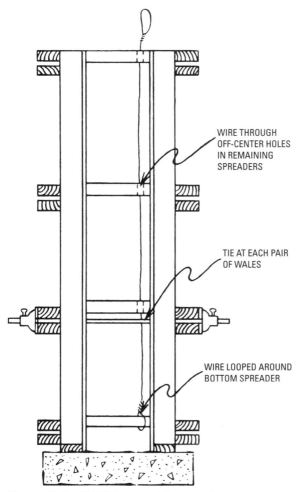

WIRE THROUGH
OFF-CENTER HOLES
IN REMAINING
SPREADERS

TIE AT EACH PAIR
OF WALES

WIRE LOOPED AROUND
BOTTOM SPREADER

Figure 3-9 Wall form with spreaders.

Because forms are removable, the form designer has certain restrictions. The designer must not only consider erection, but also stripping. Thus, if a form is designed in such a way that final-assembly nails are covered, it may be impossible to remove the form without tearing it apart and possibly damaging the partially cured concrete.

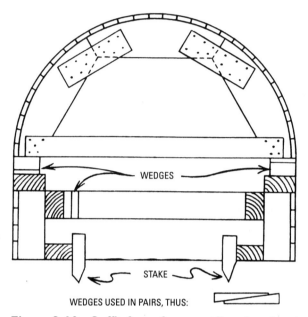

WEDGES

WEDGES USED IN PAIRS, THUS:

STAKE

Figure 3-10   Soffit form for a small arch culvert.

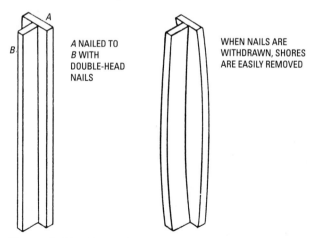

A

B

A NAILED TO
B WITH
DOUBLE-HEAD
NAILS

WHEN NAILS ARE
WITHDRAWN, SHORES
ARE EASILY REMOVED

Figure 3-11   Form for a small job.

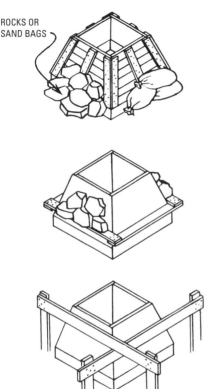

ROCKS OR
SAND BAGS

Figure 3-12 Ways of anchoring pedestal footings.

Another bad design may be one in that the form is built so that some of its members are encased in concrete and may be difficult to remove. This may result in defective work. It is often advisable to plan column forms so that they can be stripped without disturbing the forms for the beams and girders.

For easier stripping, forms can be coated with special oils. These are not always effective, however, and a number of coating compounds that work well have been developed over the years. These compounds reduce the damage to concrete when stripping is difficult or perhaps carelessly done. The use of these coatings reduces the importance of wetting formwork before placing concrete.

After stripping, forms should be carefully cleaned of all concrete before they are altered and oiled for reuse.

Figure 3-13   Cardboard box being used as a form in a concrete floor-slab pour.

Stripping Forms for Arches

Forms for arches, culverts, and tunnels generally include hinges or loose pieces that, when removed, release the form (see Figure 3-10). Shores are often set on screw jacks or wedges to simplify their removal. Screw jacks are preferable to wedges because forms held in jacks can be stripped with the least amount of hammering.

For small jobs where jacks are not available, shores that are almost self-releasing can be made. Two 2 × 6 boards are fastened into a T-shaped section (see Figure 3-11) with double-headed nails. When wedged into position, this assembly is a stiff column. After the concrete hardens, the nails are drawn and the column becomes two 2 × 6 boards, which are relatively easy to remove.

Special Forms

Forms are required for building concrete piers, pedestals, and foundations for industrial machinery. The job still involves carpentry,

Figure 3-14 Prefabricated build-up panels used for concrete forms.

because such forms differ only in detail from building forms, and usually do not have to withstand the pressures that are built up in deep wall or column forms.

Many piers are tapered upward from the footing. In these cases, it is necessary to provide a resistance to uplift because the semi-liquid concrete tends to float such forms. Once the problem is recognized, it is easily solved. Two or more horizontal planks nailed or wired to spikes will hold down most forms. Sandbags placed on ledges (see Figure 3-12) are usually enough for smaller forms.

Some industrial forms are complicated and require cast-in-place hold-down bolts for the machinery. If these bolts are not needed, however, the work is greatly simplified. Foundations can also be built with recesses into which bolts threaded on both ends can be dropped through pipelined holes. The machine can be slid onto such

Figure 3-15   Peg-and-wedge system used to connect and hold prefabricated panels together.

Figure 3-16   Circular metal concrete form.

a foundation and jacked into position without too much difficulty. If desired, the bolts can be grouted after they are in place.

Simple forms may be made by using a cardboard box filled with wet sand, gravel, or dirt (see Figure 3-13). This can be used where a slab is poured at grade level and there is a need for drains, toilets, water closets, or showers to penetrate the slab.

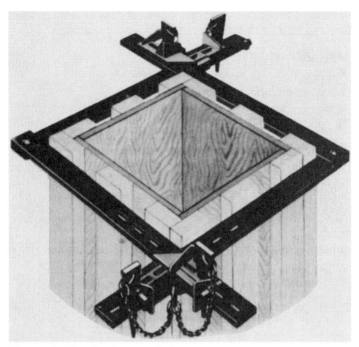

Figure 3-17   Spring-steel clamps.

Prefabricated Forms

In addition to lumber and plywood forms, there are a tremendous variety of prefabricated forms (see Figure 3-14). These panels can be made up in different widths and lengths. There is a peg-and-wedge system to hold the panels together (see Figure 3-15).

A variety of hinged forms are available, as well as circular metal ones (see Figure 3-16) and clamps (see Figure 3-17).

Summary

Concrete mixture is a semi-fluid that will take the shape of any form into which it is poured. Forms are usually made of metal or wood. Forms must be reasonably tight, rigid, and strong enough to sustain the weight of the concrete.

Most concrete construction (such as building walls) is done by using metal forms. Oiling or greasing the inside of the form before use prevents absorption of water from the concrete, which could buckle or warp the forms and weaken the concrete.

Forms may also be prefabricated. Forms should be regarded as a structure and built with economy and efficiency in mind.

Review Questions

  1. Why are concrete forms oiled or greased?

  2. What are prefabricated forms?

  3. Name two ways to economize when building forms.

  4. How much does concrete weigh per cubic foot?

  5. What does the term *tie* mean in formwork?

  6. Why should economy be factored into a decision to make a form or buy prefab?

  7. A concrete mixture is ————.

  8. How much does formwork cost in reference to the concrete work?

  9. What is a wale?

  10. Where are wedges used in formwork?

Chapter 4

Site Equipment

The proper way to set up a ladder or scaffold is very important in carpentry work. There are many applications for ladders. Being able to safely handle a ladder is most important in any carpentry work. There is always a roof to a building. It needs shingles and other preparation that needs access to locations above ground level. Scaffolding is useful in painting and putting on the siding. If the house is more than one story, you would probably include scaffolding to finish up around the inside of the house where a skylight might be mounted. For high ceilings, the scaffold is also needed for drywall work and painting. Electricians and plumbers also use ladders and scaffolds to perform their duties.

Ladders

The ladder is one of the most commonly used tools of the carpenter. Ladders are made in a number of sizes and shapes, and they serve a number of purposes in the construction business. You may use a simple stepladder, or you may need complicated scaffolding for support high above the ground or floor.

The *single straight ladder* consists of one section with two side rails and several rungs made of round dowel rods in most instances. Single ladders are available in 8- to 16-foot lengths and 18- to 20-foot lengths. The 8- to 16-foot size will have $1^3/_8$- to $2^3/_4$-inch side rails, whereas the 18- to 20-foot size has side rails up to 3 inches wide. These ladders are usually 16 inches wide both at the bottom and at the top, and do not taper (see Figure 4-1).

Push-up ladders are convenient, light, and expandable. They are made in 16-, 20-, 24-, and 28-foot lengths. The maximum working length is 3 feet shorter than the specified length of the ladder. For example, the 16-foot ladder has a working length of 13 feet, and the 28-foot ladder has a working length of 25 feet. These ladders can be secured at a certain length by metal brackets hooked over the side rails with an automatic catch (see Figure 4-2).

Most *metal extension ladders* are made of aluminum, which makes them lighter than the same size of wooden ladder. The metal extension ladder (see Figure 4-3) operates with a rope and pulley. An automatic catch holds the extended ladder in place against the rung of the bottom ladder. A 14-foot ladder of this type weighs about $17^1/_2$ pounds. Metal extension ladders are available in total extended lengths of 16, 20, 24, 28, 32, 36, and 40 feet. All of them

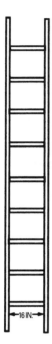

Figure 4-1 Standard straight wooden ladder. *(Courtesy of Waco Scaffolding and Equipment)*

Figure 4-2 Push-up wooden extension ladder.

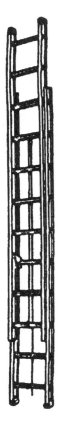

Figure 4-3 Aluminum flat-step extension ladder.
(Courtesy of Waco Scaffolding and Equipment)

must overlap at least 3 feet, except the 40-foot size, which requires a 4-foot overlap to safely support the weight of a person.

Fiberglass ladders are available for those who must work near electrical lines. It is important to choose a ladder that does not conduct current if you are going to work near electrical lines.

Magnesium ladders are available in both stepladder and extension-ladder configurations. Magnesium has the advantage of being a lightweight metal.

Stepladders, as previously mentioned, can be made of wood, aluminum, or magnesium. They may be made of pine (see Figure 4-4), with flat front legs braced with metal brackets. A metal tie rod may (or may not, depending on price) be inserted under each step to hold the side rails steady and to prevent wobble as the ladder ages

Figure 4-4 Southern pine commercial stepladder. *(Courtesy of Waco Scaffolding and Equipment)*

with use. The dowels are usually made of hickory or ash, while the side-rails and steps are made of pine. Hemlock is used for side rails and steps in some wooden stepladders. The Occupational Safety and Health Administration (OSHA) rates stepladders as commercial Type I, Type II, or Type III. It also rates them as industrial or household types with a I, II, or III rating.

Light household stepladders have a duty rating of 225 pounds. That means the step will hold 225 pounds at the center. In most instances, the weight supported by the ladder will be no greater than this. To be on the safe side, however, the center of the step must be able to support four times that weight (in others words, 900 pounds). The ladder illustrated in Figure 4-5 has braced bottom and top steps and uses vinyl shoes. These ladders come in heights of 3, 4, 5, and 6 feet.

The common *trestle* has no extension and resembles a triangle (see Figure 4-6). The cross-members are 1-inch × 2-inch strips of oak. Trestles are made in heights of 6, 8, 10, 12, 14, and 16 feet. The distance between the side rails measures 15.25 inches inside. A trestle weighs about 5 pounds per foot. Planks or platforms can be suspended between two trestles to provide a runway. However, this may not be high enough, so the trestle is also available with extensions. The 6-foot trestle extends to $9^1/_2$ feet; the 8-foot size goes to $13^1/_2$ feet; and the 10-foot size will extend to $16^1/_2$ feet. The 12-foot ladder is will extend to $20^1/_2$ feet. This one is a little heavier, since it has an extension as part of the package. It will weigh

Pail shelf with
rag rail and tool
holders.

Figure 4-5 Aluminum stepladder with vinyl shoes. *(Courtesy of Waco Scaffolding and Equipment)*

Figure 4-6 Common trestle.

about 7 pounds per foot (see Figure 4-7). The extensions can be placed so that the platform can be mounted between two trestles to support one or two people. The strength of the platform is important for safety reasons. Figure 4-8 shows what can happen when two stepladders are used to support an ordinary plank. The ladders with the Stinson plank provide a more secure platform that does not sag. A sagging plank can cause a person to become unbalanced and fall.

The *New York extension trestle* has a special locking device that holds the middle ladder in such a way as to eliminate all wobble. This device is so designed that any weight applied on the ladder will

Figure 4-7 Heavy-duty extension trestle.
(Courtesy of Waco Scaffolding and Equipment)

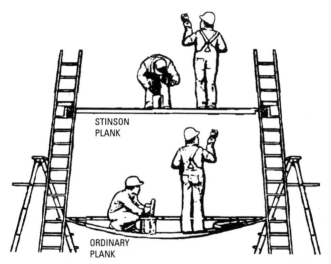

STINSON
PLANK

ORDINARY
PLANK

Figure 4-8 The plank makes a difference. *(Courtesy of Waco Scaffolding and Equipment)*

increase the grip on the lock. The middle ladder has a rung spacing of 8 inches to permit adjustment at the upper levels. This is the most expensive of the trestle ladders. It comes in lengths of 6, 8, 10, 12, 14, and 16 feet. The 6-foot ladder extends to 10 feet; the 8-foot to 14 feet; the 10-foot to 18 feet; the 12-foot to 22 feet; the 14-foot to 26 feet; and the 16-foot to 30 feet.

Figure 4-9 shows how a fiberglass trestle ladder is used to support a scaffold plank.

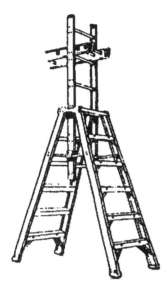

Figure 4-9 Fiberglass extension trestle ladder. *(Courtesy of Waco Scaffolding and Equipment)*

Setting Up a Ladder

Raising a ladder can be a two-person job. In fact, a heavier ladder should have two workers on it. However, if you have a single ladder that's not too heavy, you can place the end of the ladder against the house or some obstruction and walk it up one rung at a time (see Figure 4-10). Keep in mind that the top of the ladder should be placed against the house. The bottom of the ladder will have a distance from the house of one-fourth the length of the ladder. Some safety tips will be given later in this chapter as to what angles to use, how much overlap to allow for extension ladders, and how to make sure the ladder does not slip after it has been placed. The main thing is to get the right ladder for the job at hand. A ladder too long can cause as much trouble as one that is too short. The strength of the ladder is also important, since it must support one, two, or more people as well as building materials.

Ladder Shoes

To keep the ladder from slipping while you are on it, anchor the bottoms of the side rails with *ladder shoes.* These shoes fit over the ends of the side rails. A number of types are available. The rubber boot keeps the ladder from slipping as a person climbs. The rubber is nonconductive and useful in corrosive atmospheres (see Figure 4-11).

Figure 4-10   Walking a ladder up.

Figure 4-11   Rubber-boot ladder shoe. *(Courtesy of Waco Scaffolding and Equipment)*

The *universal-safety ladder shoe* tilts and provides protection on all kinds of surfaces. It is a dual-purpose shoe, as it is equipped with steel spikes and with suction-grip composition treads (see Figure 4-12).

The *steel-spur wheel shoe* (see Figure 4-13) has steel points. When it becomes worn, the wheel can be turned to expose new points. This shoe is used extensively by public utility companies and industries.

Stepladder shoes (see Figure 4-14) can be added by inserting a bolt through the side rails or legs of the ladder. They can be easily

Figure 4-12 Universal-safety ladder shoe.
(Courtesy of Waco Scaffolding and Equipment)

Figure 4-13 Steel-spur wheel shoe. *(Courtesy of Waco Scaffolding and Equipment)*

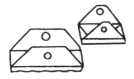

Figure 4-14 Stepladder shoes. *(Courtesy of Waco Scaffolding and Equipment)*

adjusted to fit the angle of the terrain on which the ladder is being used. Composition soles are placed on the bottom of the shoes to prevent slipping.

Ladder Accessories

Ladders, like everything else these days, have accessories that can be added to make them adaptable to almost any purpose. Figure 4-15 shows several ladder accessories: a *house pad* can be added to prevent sliding along a wall; a *pole strap* can be added to make the ladder secure against a pole when necessary. Other safety devices can be ordered to fit any ladder.

Special Products

The *aluminum ladder jack* comes in handy when you need to put a plank between two ladders to serve as a platform. The aluminum ladder jack (see Figure 4-16) has an adjustable arm with a positive lock. Adjustments are every inch, so it can be used on either side of the ladder. It can be installed on ladders with roof hooks, enabling the user to work on the roof or the face of a dormer. It

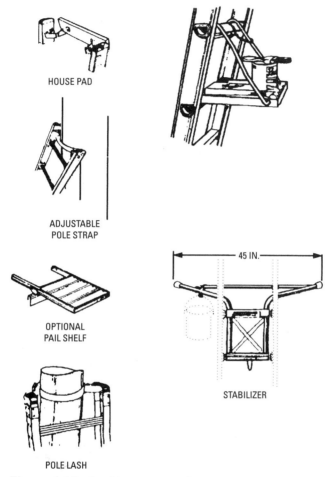

HOUSE PAD

ADJUSTABLE
POLE STRAP

OPTIONAL
PAIL SHELF

45 IN.

STABILIZER

POLE LASH

Figure 4-15 Ladder accessories. *(Courtesy of Waco Scaffolding and Equipment)*

accommodates forms up to 18 inches wide, but folds to 6 inches deep for storage.

The *side-rail ladder jack* (see Figure 4-17) has a four-point suspension that engages the side rails of the ladder. It fits ladders with round or D-shaped rungs. It can be used on either side of the ladder, so 12-inch planks fit over the jack. A larger type will take 20-inch-wide aluminum planks, but it has oversized hooks that span double rails.

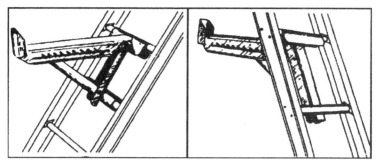

Figure 4-16 Ladder jack. *(Courtesy of Waco Scaffolding and Equipment)*

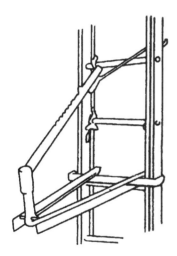

Figure 4-17 Side-rail ladder jack.

(Courtesy of Waco Scaffolding and Equipment)

The *ladder hook* comes in handy for a number of jobs. It attaches to the ladder quickly and fits over an edge or the top of a wall. The screw device on the spine clamps over one rung of the ladder, and the bottom curve fits against a lower rung (see Figure 4-18).

Telescoping extension planks are light, durable, and convenient for light interior use by one person (see Figure 4-19). They come in two and three sections only. The outside width is 12 inches, and the length can be 6, 8, or 10 feet.

A *truck caddy rack* comes in handy for transporting ladders from the work site to storage and back to the work site. The caddy (see Figure 4-20) fits on a pickup truck. The rack has only four basic

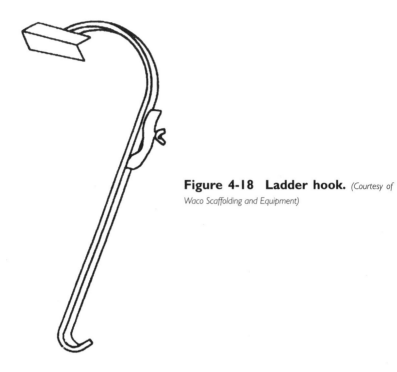

Figure 4-18 Ladder hook. *(Courtesy of Waco Scaffolding and Equipment)*

Figure 4-19 Telescoping extension planks. *(Courtesy of Waco Scaffolding and Equipment)*

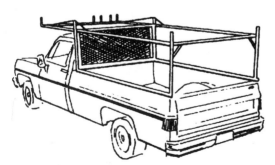

Figure 4-20 Truck caddy rack for ladders and planks. *(Courtesy of Waco Scaffolding and Equipment)*

parts and bolts together easily. This caddy allows you to secure the ladders safely.

Ladder Safety

Ladders are only as safe as the user makes them. You can make them accident-proof with a little effort. The best bet is to use ladders that meet all local code requirements. OSHA has established standards for ladders, which will be marked on the side rail so that you know into which classification your ladder fits. Wooden ladders should not be painted, but you can coat them with varnish or another clear finish. The reason for not painting a ladder is that you must be able to inspect the wood for cracks and defects in the grain.

Keep the screws and bolts tightened so that the braces and all other parts of the ladder remain in top condition at all times. Remember, your life may depend on the condition of the ladder.

Make sure you use the proper ladder for the job. The ladder should extend at least 3 feet above the roofline if you plan to use it to climb onto the roof. This extension will give you something to grasp as you get on and off the ladder.

The proper anchoring of the ladder is very important. It isn't a nice feeling to get on the ladder and have it start to rock back and forth or slide along the wall. The correct angle for the ladder is 75 degrees with the ground. The space from the foot of the ladder to the wall should be one-fourth the length of the ladder.

Of course, you should face the ladder as you climb up and down. Don't lean too far to the left or right while working on the ladder. Keep one hand on the side rail and the other hand on the rungs as you climb or descend the ladder. If you are carrying paint or tools in one hand, use your other hand to hold on to the side rail as you climb. Put your feet firmly on the rungs, and make sure your shoes and the rungs are free from grease, mud, and other substances. Keep your feet and the ladder clean for safety's sake. Just keep one hand on the ladder at all times.

Make sure the ladder is properly shimmed so it fits snugly into the ground. Ladder shoes can be helpful in most instances. You can adjust them to fit any angle, and they have nonskid material on the soles. In some cases, you may need to place a board under the ladder to level it before climbing.

Keep in mind that it takes two people to get longer extension ladders in place. Don't try to place a long ladder by yourself. Ask for help. If you do it alone, you risk breaking windows and straining muscles. However, do not allow two people to stand on a ladder at the same time.

Scaffolding

Scaffolding allows you access to high places quickly and efficiently. It can be movable, or it can be held in place with permanent brackets or nailed to boards. In this chapter, we cover primarily scaffolding that can be assembled and torn down for easy storage or transport.

A number of features should be examined in the interest of safety. State and federal regulations apply to all scaffolding, so the manufacturer is aware of the limitations of a specific arrangement. Follow the manufacturer's recommendations for a safe tower or support platform.

In this chapter we will take a quick look at scaffolding and its basic components. As you view the illustrations, you will begin to see what scaffolding is, how it is used, and what special advantages this type of construction offers.

Scaffolding Components

Scaffolding has two basic parts: *end frames* and *cross braces* (see Figure 4-21). Several basic sections can be joined together vertically to form a tower, or they can be joined side-by-side to make a run (Figure 4-22). Alternatively, you can use a combination of the tower and the run. A device called the *Speed lock* secures braces to end frames (see Figure 4-23). It will hold one brace (which is necessary in towers) or two (which are needed in runs of scaffolding).

The *coupling pin* (see Figure 4-24) is used to stack one frame on top of another when you are building a vertical scaffold. The coupling pin is designed to permit building up frames one leg at a time. When building rolling towers, or when uplift of frames could occur, frames should be locked together with toggle pins or with bolts and nuts.

End frames for sectional scaffolding come in two basic widths: the standard 5-foot-wide end frames in various shapes, and the $27^5/_8$-inch-wide narrow frames (see Figure 4-25). The height selected depends on trade or union preferences and the nature of the work. The 6-foot, 7-inch-high frames are practical for higher scaffolds, because fewer units need to be assembled or dismantled for a given height and because their higher overhead clearance makes them easy to walk through. Extension frames and end frames allow you to build scaffolding to any convenient height and to tier or contour it to suit ceiling contours and other projections. On exterior walls, the walk-through frame is frequently used. When combined with side brackets, it allows maximum clearance and freedom of movement.

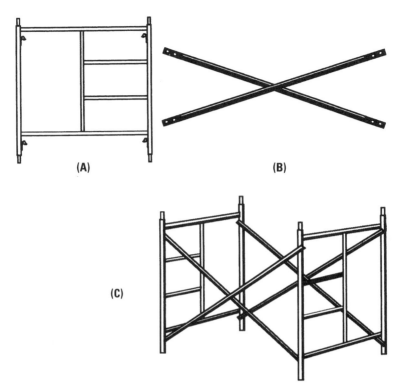

Figure 4-21 **(A) End frames and (B) cross braces comprise (C) scaffolding.** *(Courtesy of Waco Scaffolding and Equipment)*

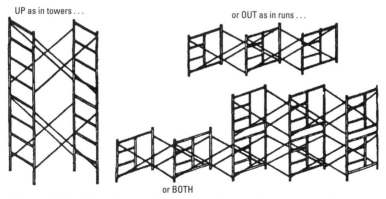

UP as in towers . . .

or OUT as in runs . . .

or BOTH

Figure 4-22 **Putting end frames and cross braces together.** *(Courtesy of Waco Scaffolding and Equipment)*

(A) Lift Speedlock. Retaining pin allows it to be raised far enough to permit attachment of brace.

(C)

(B) Brace is attached by slipping end over the Speedlock stud. Release Speedlock and brace is secured in place.

Figure 4-23 Speed lock secures braces to the end frames. *(Courtesy of Waco Scaffolding and Equipment)*

Note the combination walk-through frame in Figure 4-26. The *guardrail post* slips over the top of the frame, and is attached to provide a safe working area on top of the frame. Figure 4-27 shows the guardrail and toe-board assemblies used with standard scaffolding. Guardrail post, guardrails, and toe-board assemblies should be used on all types of scaffolding installations (both rolling and stationary) according to applicable federal and state safety regulations. Occasionally, guardrails must be provided for unusual spacing. They can

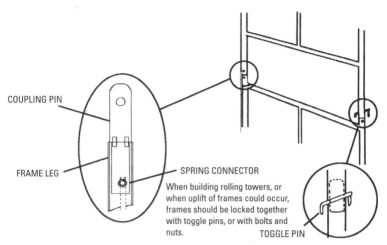

COUPLING PIN

FRAME LEG

SPRING CONNECTOR
When building rolling towers, or when uplift of frames could occur, frames should be locked together with toggle pins, or with bolts and nuts.

TOGGLE PIN

Figure 4-24   A coupling pin is used to stack one frame on top of another. (Courtesy of Waco Scaffolding and Equipment)

be held by using a *nailing plate* that goes around a guard post and to that timber can be nailed.

Cross-braces come in two types and various lengths. There are *single-hole braces* and *double-hole braces*. The length of the braces will depend on the length of the planks to be used, the weight the scaffold must support, and the area to be covered. Double-hole braces can be used on different-sized frames to obtain the same frame spacing (see Figure 4-28). Straddle braces are available in lengths of 7 or 10 feet. These permit scaffolding to be erected over obstructions. The purpose is to straddle furniture, machinery, materials, or whatever is in the way. They will allow worker traffic under the scaffolding (see Figure 4-29).

After determining how high you will go with the scaffold and the preferred height of the frames, select the number of frames and braces needed for the length of the run by referring to Table 4-1.

Figure 4-30 shows the *hoist standard*. It can be pinned to the top of any frame. It is an easy way to provide an easy method for moving up to 100 pounds of material to the top of the scaffold. The hoist standard is slipped over the coupling pin and pinned to the leg. The head is complete with a 12-inch well wheel that holds rope up to 1-inch thick. This head swivels so that materials may be swung over the scaffold platforms.

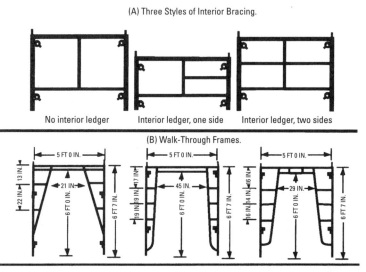

(A) Three Styles of Interior Bracing.

No interior ledger   Interior ledger, one side   Interior ledger, two sides

(B) Walk-Through Frames.

Note: All vertical dimensions indicate leg height. Leg extend 1 inch above ledger on all frames.

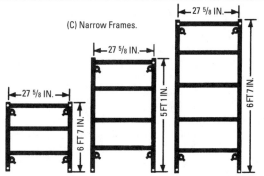

(C) Narrow Frames.

Figure 4-25   Different sizes and shapes of frames. *(Courtesy of Waco Scaffolding and Equipment)*

Figure 4-31 shows a pair of *platform supports* bolted to a piece of plywood to make a convenient platform for a rolling tower.

Figure 4-32 shows a tower of the rolling type used by painters. Rolling towers can be made from standard end frames fitted with horizontal braces and casters. The *casters* (see Figure 4-33) are available in 5- and 8-inch diameters and are equipped with a brake permitting the user to lock the wheel in position. When assembling

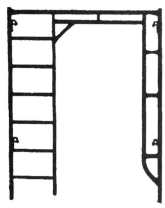

Figure 4-26 Combination walk-through frame. *(Courtesy of Waco Scaffolding and Equipment)*

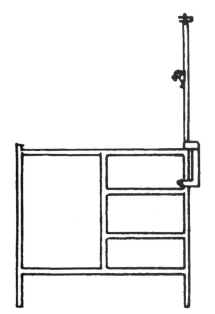

Figure 4-27 Guard rail. *(Courtesy of Waco Scaffolding and Equipment)*

Figure 4-28   Cross braces. *(Courtesy of Waco Scaffolding and Equipment)*

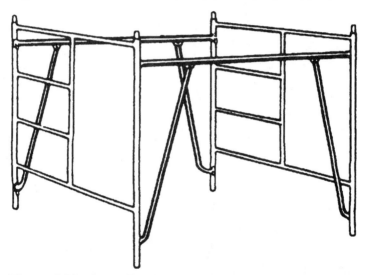

Figure 4-29   Straddle braces. *(Courtesy of Waco Scaffolding and Equipment)*

a rolling tower, always use horizontal braces on the frame section. This prevents the tower from racking (getting out of square). On rolling towers, horizontal braces must be used at the bottom and at every 20-foot height measured from the rolling surface (see Figure 4-34).

Scaffolding safety rules are reprinted here with the permission of the Scaffolding and Shoring Institute.

Scaffolding Safety Rules
Following are some commonsense rules, as recommended by the Scaffolding and Shoring Institute, designed to promote safety in the use of steel scaffolding. These rules are illustrative and suggestive

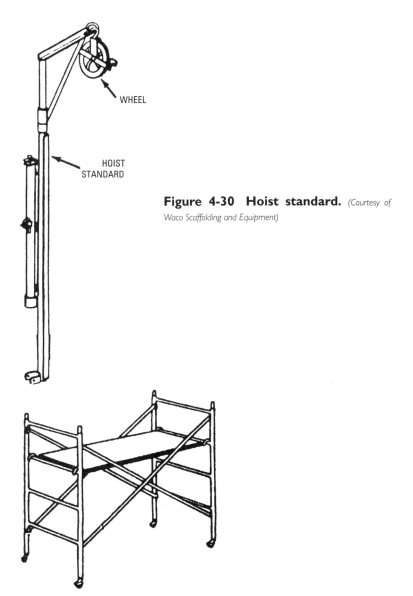

WHEEL

HOIST
STANDARD

Figure 4-30 Hoist standard. *(Courtesy of Waco Scaffolding and Equipment)*

Figure 4-31 Platform supports bolted to a piece of plywood.
(Courtesy of Waco Scaffolding and Equipment)

Table 4-1 Scaffolding Selection*

No. of Frames High	4-Foot Frames	4½-Foot Frames	5-Foot Frames	6½-Foot Frames
1	4 feet	4½ feet	5 feet	6½ feet
2	8 feet	9 feet	10 feet	13 feet
3	12 feet	13½ feet	15 feet	19½ feet
4	16 feet	18 feet	20 feet	26 feet
5	20 feet	22½ feet	25 feet	32½ feet
6	24 feet	27 feet	30 feet	39 feet
7	28 feet	31½ feet	35 feet	45½ feet
8	32 feet	36 feet	40 feet	52 feet
9	36 feet	40½ feet	45 feet	58½ feet
10	40 feet	45 feet	50 feet	65 feet
11	44 feet	49½ feet	55 feet	71½ feet
12	48 feet	54 feet	60 feet	78 feet
13	52 feet	58½ feet	65 feet	84½ feet
14	56 feet	63 feet	70 feet	91 feet

Note: Allowable plank span and scaffold loading must be in accordance with federal and state regulations.
*Seven-foot spacing assumed. Bold figures indicate number of braces.

only, and are intended to deal only with some of the many practices and conditions encountered in the use of scaffolding. These rules do not purport to be all-inclusive or to supplant or replace other additional safety and precautionary measures to cover usual or unusual conditions. They are not intended to conflict with or supersede any state, local, or federal statute or regulation; reference to such specific provisions should be made by the user. Reprinting of these rules does not imply approval by the Institute or indicate membership in the Institute.

Table 4-1 (continued)

7-Foot Frames	14-Foot Frames	21-Foot Frames	28-Foot Frames	35-Foot Frames	42-Foot Frames
2	3	4	5	6	7
2	4	6	8	10	12
4	6	8	10	12	14
4	8	12	16	20	24
6	9	12	15	18	21
6	12	18	24	30	36
8	12	16	20	24	28
8	16	24	32	40	48
10	15	20	25	30	35
10	20	30	40	50	60
12	18	24	30	36	42
12	24	36	48	60	72
14	21	28	35	42	49
14	28	42	56	70	84
16	24	32	40	48	56
16	32	48	64	80	96
18	26	36	45	54	63
18	36	54	72	90	108
20	30	40	50	60	70
20	40	60	80	100	120
22	33	44	55	66	77
22	44	66	88	110	132
24	36	48	60	72	84
24	48	72	96	120	144
26	39	52	65	78	91
26	52	78	104	130	156
28	42	56	70	84	98
28	56	84	112	140	168

- *Post these scaffolding safety rules* in a conspicuous place and be sure that all persons who erect, dismantle, or use scaffolding are aware of them.
- *Follow all state, local, and federal codes, ordinances, and regulations* as they pertain to scaffolding.
- *Inspect all equipment before using.* Never use any equipment that is damaged or deteriorated in any way.
- *Keep all equipment in good repair.* Avoid using rusted equipment (the strength of rusted equipment is not known).
- *Inspect erected scaffolds regularly.* Be sure that they are maintained in safe condition.

Table 4-1 Scaffolding Selection* (*continued*)

No. of Frames High	49-Foot Frames	56-Foot Frames	63-Foot Frames	70-Foot Frames	77-Foot Frames	84-Foot Frames
1	8	9	10	11	12	13
	14	**16**	**18**	**20**	**22**	**24**
2	16	18	20	22	24	26
	28	**32**	**36**	**40**	**44**	**48**
3	24	27	30	33	36	39
	42	**48**	**54**	**60**	**66**	**72**
4	32	36	40	44	48	52
	56	**64**	**72**	**80**	**88**	**96**
5	40	45	50	55	60	65
	70	**80**	**90**	**100**	**110**	**120**
6	48	54	60	66	72	78
	84	**96**	**108**	**120**	**132**	**144**
7	56	63	70	77	84	91
	98	**112**	**126**	**140**	**154**	**168**
8	64	72	80	88	96	104
	112	**128**	**144**	**160**	**176**	**192**
9	72	81	90	99	108	117
	126	**144**	**162**	**180**	**198**	**216**
10	80	90	100	110	120	130
	140	**160**	**180**	**200**	**220**	**240**
11	88	99	110	121	132	143
	154	**176**	**198**	**220**	**242**	**264**
12	96	108	120	132	144	156
	168	**192**	**216**	**240**	**264**	**288**
13	104	117	130	143	156	169
	182	**208**	**234**	**260**	**286**	**312**
14	112	126	140	154	168	182
	196	**224**	**252**	**280**	**308**	**336**

*Seven-foot spacing assumed. Bold figures indicate number of braces.

- *Consult your scaffolding supplier when in doubt.* Scaffolding is the supplier's business. *Never take chances.*
- *Provide adequate sills* for scaffold posts, and use base plates.
- *Use adjusting screws* instead of blocking to adjust to uneven grade conditions.
- *Plumb and level all scaffolds* as the erection proceeds. Do not force braces to fit. Level the scaffold until proper fit can be made easily.
- *Fasten all braces securely.*

Table 4-1 (*continued*)

91-Foot Frames	98-Foot Frames	105-Foot Frames	112-Foot Frames	119-Foot Frames	126-Foot Frames	133-Foot Frames
14	15	16	17	18	19	20
26	28	30	32	34	36	38
28	30	32	34	36	38	40
52	56	60	64	68	72	76
42	45	48	51	54	57	60
78	84	90	96	102	108	114
56	60	64	68	72	76	80
104	112	120	128	136	144	152
70	75	80	85	90	95	100
130	140	150	160	170	180	190
84	90	96	102	108	114	120
156	168	180	192	204	216	228
98	105	112	119	126	133	140
182	196	210	224	238	252	266
112	120	128	136	144	152	160
208	224	240	256	272	288	304
126	135	144	153	162	171	180
234	252	270	288	306	324	342
140	150	160	170	180	190	200
260	280	300	320	340	360	380
154	165	176	187	198	209	220
286	308	330	352	374	396	414
168	180	192	204	216	228	240
312	336	360	384	408	432	456
182	195	208	221	234	247	260
338	364	390	416	442	468	496
196	210	224	238	252	266	280
364	392	420	448	476	504	532

- *Do not climb cross braces.* An access (climbing) ladder, access steps, a frame that is designed to be climbed, or equivalent safe access to the scaffold shall be used.
- *On wall scaffolds place and maintain anchors* securely between structure and scaffold at least every 30 feet of length and 25 feet of height.
- *When scaffolds are to be partially or fully enclosed,* specific precautions must be taken to ensure frequency and adequacy of ties attaching the scaffolding to the building because of increased load conditions resulting from effects of wind and

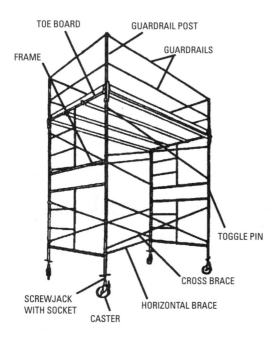

TOE BOARD
GUARDRAIL POST
GUARDRAILS
FRAME
TOGGLE PIN
CROSS BRACE
SCREWJACK WITH SOCKET
HORIZONTAL BRACE
CASTER

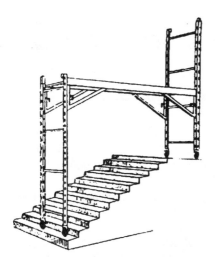

Figure 4-32   Rolling scaffolds. *(Courtesy of Waco Scaffolding and Equipment)*

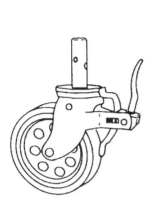

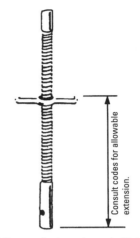

(A) Casters specially designed for use with rolling scaffolds. They are available in 5-inch and 8-inch diameters and are equipped with a brake permitting the user to lock the wheel in position.

(B) Casters wheels can be used with this adjustment screw, which allows adjustment of tower height.

Figure 4-33   Caster and adjustment screw for casters. *(Courtesy of Waco Scaffolding and Equipment)*

Figure 4-34   Note the horizontal braces on the frame section to prevent racking. *(Courtesy of Waco Scaffolding and Equipment)*

weather. The scaffolding components to which the ties are attached must also be checked for additional loads.

- *Free-standing scaffold towers must be restrained from tipping.* Use either guying or other means.
- *Equip all planked or staged areas* with guide rails, mid-rails, and toe boards along all open sides and ends of scaffold platforms.
- *Power lines near scaffolds* are dangerous. Use caution and consult the power service company for advice.
- *Do not use ladders or makeshift devices on top of scaffolds to increase height.*
- *Do not overload scaffolds.*

When using planking, keep the following in mind:

- Use only lumber that has been properly inspected and graded for scaffold planks.
- Planking shall have at least 12 inches of overlap and extend 6 inches beyond center of support, or be cleated at both ends to prevent sliding off supports.
- Fabricated scaffold planks and platforms (unless cleated or restrained by hooks) shall extend over their end supports not less than 6 inches or more than 12 inches.
- Secure plank to scaffold when necessary.

For rolling scaffold, the following additional rules apply:

- *Do not ride rolling scaffolds.*
- *Secure or remove all material and equipment* from platform before moving scaffold.
- *Caster brakes must be applied* whenever scaffolds are not being moved.
- *Casters with plain stems* shall be attached to the panel or adjustment screw by pins or other suitable means.
- *Do not attempt to move a rolling scaffold without sufficient help.* Watch out for holes in the floor and overhead obstructions.
- *Do not extend adjusting screws on rolling scaffolds more than 12 inches.*

- *Use horizontal diagonal bracing* near the bottom and at 20-foot intervals measured from the rolling surface.

- *Do not use brackets on rolling scaffold* without consideration of overturning effect.

- *The working platform height of a rolling scaffold* must not exceed four times the smallest base dimension unless guyed or otherwise stabilized.

For putlogs and trusses, the following additional rules apply:

- *Do not cantilever or extend putlogs or trusses* as side brackets without thorough consideration for loads to be applied.

- *Putlogs and trusses should extend at least 6 inches beyond point of support.*

- *Place proper bracing between putlogs or trusses* when the span of the putlog or truss is more than 12 feet.

- *All brackets* shall be seated correctly with side brackets parallel to the frames and end brackets at 90° to the frames. Brackets shall not be bent or twisted from normal position. Brackets (except mobile brackets designed to carry materials) are to be used as work platforms only and shall not be used for storage of material or equipment.

- *All scaffolding accessories* shall be used and installed in accordance with the manufacturer's recommended procedure. Accessories shall not be altered in the field. Scaffolds, frames, and their components manufactured by different companies shall not be intermixed.

Summary

Ladders and scaffolding provide access for work high above the ground. Step, single, and extension ladders are suitable for use by single workers for tasks of short duration. Ladders are made of wood, fiberglass, or metal. Ladders have a variety of attachments that help stabilize the ladder at the top and bottom. Heavier ladders should be raised by two workers.

Scaffolding provides a larger access area for longer jobs that require heavy materials at hand. Modular end frame, bracing units, and planks can be built into a variety of temporary structures. Accessories for hoisting and extensions are part of the system. Be sure to practice good safety when using ladders and scaffolding. Follow standard rules and guidelines.

Review Questions

1. What type of ladder should be used around electrical wires?

2. Why shouldn't wooden ladders be painted?

3. When setting a ladder, how far above the roofline should it extend?

4. How is a scaffold hoist installed and used?

5. Can ladders be used on top of scaffolding to increase the height?

6. Push-up ladders are convenient, light and _____.

7. What type of ladder is often used for interior painting?

8. What type of ladder is used frequently in outside siding work?

9. Explain how safety on ladders and scaffolds is important to a carpenter.

10. Why would an electrician need a ladder on the job?

Chapter 5

Concrete-Block Construction

Concrete blocks (or *cement blocks*, as they are also known) provide suitable building units. By the use of standard-size hollow blocks, walls can be erected at a very reasonable cost and are durable, light in weight, fire-resistant, and able to carry heavy loads.

The term *concrete masonry* is applied to building units molded from concrete and laid by masons in a wall (see Figures 5-1 and 5-2). The units are made of Portland cement, water, and suitable aggregates (such as sand, gravel, crushed stone, cinders, burned shale, or processed slag). The units are laid in a bed of cement mortar.

Figure 5-1   Two-core masonry block being installed. *(Courtesy of Bilco Company.)*

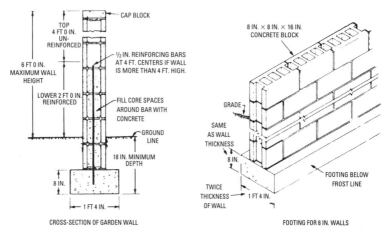

CROSS-SECTION OF GARDEN WALL   FOOTING FOR 8 IN. WALLS

Figure 5-2   Cross-section and view of a simple block wall. Vertical reinforcement rods are placed in the hollow cores at various intervals.

Block Building Materials

The two key materials used in concrete block construction are the standard blocks themselves (available in various sizes and shapes) and the mortar used to bind the blocks together.

Standard Masonry Units

Concrete blocks are available in many sizes and shapes (see Figure 5-3). They are all sized based on multiples of 4 inches. The fractional dimensions shown allow for the mortar (see Figure 5-4). Some concrete blocks are poured concrete made of standard cement, sand, and aggregate. An 8-inch × 8-inch × 16-inch block weighs about 40–50 pounds. Some use lighter natural aggregates (such as volcanic cinders or pumice), while some are manufactured aggregates (such as slag, clay, or shale). These blocks weigh 25–35 pounds.

In addition to the hollow-core types shown, concrete blocks are available in solid forms. In some areas, they are available in sizes other than those shown. Many of the same type have half the height, normally 4 inches (actually $3^{5}/_{8}$ inches to allow for mortar). The 8 × 8 × 16 stretcher is most frequently used. It is the main block in building a yard wall or a building wall. Corner or bull-nose blocks with flat, finished ends are used at the corners of walls. Others have special detents for windowsills, lintels, and doorjambs.

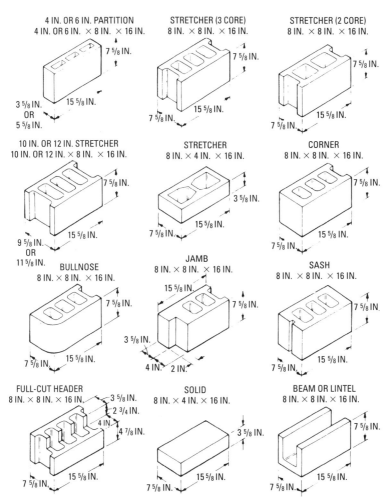

Figure 5-3 Standard sizes and shapes of concrete blocks.

Compressive strength is a function of the face thickness. Concrete blocks vary in thickness of face, depending on whether they are to be used for nonload-bearing walls (such as yard walls) or load-bearing walls (such as for buildings).

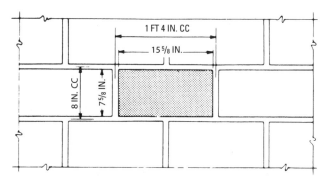

Figure 5-4   Block size to allow for mortar joints.

Mortar

Mortar bonds the masonry units together to form a strong, durable wall. The mortar must be chemically stable and resist rain penetration and damage by freezing and thawing. Mortar must have sufficient strength to carry all loads applied to the wall for the life of the building with a minimum of maintenance.

Mortar is widely used in home construction for all types of masonry walls. Masonry cement eliminates the need to stockpile and handle extra material and reduces the chance of improper on-the-job proportioning. Consistent mortar color in successive batches is easy to obtain when using masonry cement.

Water is added until the mortar is plastic and handles well under a trowel. Mortar should be machine-mixed whenever practical. Masonry cement contains an air-entraining agent that causes the formation of tiny air bubbles in the mortar. These bubbles make the mortar more workable when plastic, slow the absorption of water by the masonry unit, and reduce the possibility of weather damage.

Concrete masonry walls that are subject to average loading and exposure should be laid up with mortar. The mortar is composed of:

- One volume of *masonry* cement and 2–3 volumes of damp, loose mortar sand

or,

- One volume of *Portland* cement, 1–1¼ volumes of hydrated lime or lime putty, and 4–6 volumes of damp, loose mortar sand

Enough water is added to produce a workable mixture.

Concrete masonry walls and isolated piers are sometimes subject to severe conditions. These conditions are extremely heavy loads, violent winds, earthquakes, severe frost action, or other conditions requiring extra strength. They should be laid up with mortar composed of:

* One volume of *masonry* cement, plus 1 volume of *Portland* cement, and 4–6 volumes of damp, loose mortar sand

or,

* One volume of *Portland* cement, 2–3 volumes of damp, loose mortar sand, and up to $1/4$ volume of hydrated lime or lime putty

Enough water is added to produce a workable mixture.

The type of mortar in bearing walls in heavily loaded buildings is properly governed by the loading. Allowable working loads are commonly 70 pounds per square inch (psi) of gross wall area when laid in 1:1:6 (1 volume of Portland cement, 1 volume of lime putty or hydrated lime, and 6 volumes of damp, loose mortar sand).

On small jobs, masonry cement can be purchased and the mortar mixed with 1 part masonry cement to 3 parts sand (mix dry before adding water). A mortar mix can also be purchased to which only water is added. When a mortar of maximum strength is desired for use in load-bearing walls, or walls subjected to heavy pressure, freezing, and thawing, a mortar made of 1 part Portland cement, 1 part masonry cement, and not more than 6 parts sand is recommended.

Block Building Methods

There are some rather complicated problems for the first-time block layer. Certain procedures must be followed, and the block-laying done in sequence to allow for the addition of windows, doors and floor joists. Basic block-laying practices are discussed in the following sections. Identification of the various blocks and their placement and use is very important. The ratio of cement to sand is also important. Adding water to the mixture can be a matter of experience with the mixer and demands of the mason.

Basic Block-Laying

The usual practice is to place the mortar in two separate strips, both for the horizontal (or *bed joints*) and for the vertical (or *end joints*), as shown in Figure 5-5. The mortar is applied only on the face shells

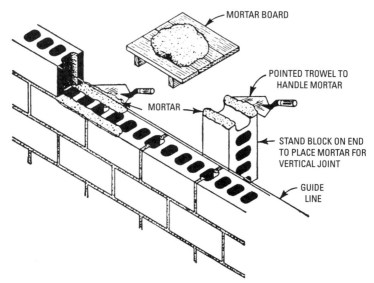

Figure 5-5 **The usual practice in applying mortar to concrete blocks.**

of the block. This is known as *face-shell bedding*. The air spaces thus formed between the inner and outer strips of mortar will help produce a dry wall.

Applying Mortar to Blocks

Masons often stand the block on end and apply mortar for the end joint. Sufficient mortar is put on to make sure that the joint will be well filled (that is, will have no air spaces). Some masons apply mortar on the end of the block previously laid, as well as on the end of the block to be laid next to it, to make certain that the vertical joint will be completely filled.

Placing and Setting Blocks

In placing (see Figure 5-6), the block that has mortar applied to one end is picked up and shoved firmly against the block previously placed. Note that the mortar is already in place in the bed or horizontal joints.

Mortar squeezed out of the joints is carefully cut off with the trowel and either applied to the other end of the block or thrown back onto the mortarboard for use later. The blocks are laid to touch the guideline and are tapped with the trowel to get them straight and

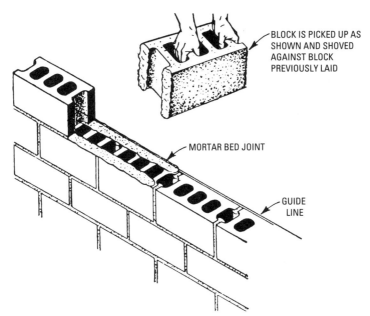

BLOCK IS PICKED UP AS
SHOWN AND SHOVED
AGAINST BLOCK
PREVIOUSLY LAID

MORTAR BED JOINT

GUIDE
LINE

Figure 5-6 The common method of picking up and setting concrete block.

level (see Figure 5-7). In a well-constructed wall, mortar joints will average ³/₈ inch thick.

Tooling Mortar Joints

When the mortar has become firm, the joints are tooled, or finished. Various tools are used for this purpose. A rounded or V-shaped *steel jointer* is the tool most commonly used. Tooling compresses the mortar in the joints, forcing it up tightly against the edges of the block, and leaves the joints smooth and watertight. Some joints may be cut off flush and struck with the trowel.

The concave and V joints are best for most work (see Figure 5-8). While the raked and the extruded styles are recommended for interior walls only, they may be used outdoors in warm climates where rain and freezing weather are at a minimum. In climates where freezing occurs, it is important that no joint permits water to collect.

Laying Blocks at Corners

In laying up corners (building the leads), place a taut line all the way around the foundation, with the ends of the string tied together. It

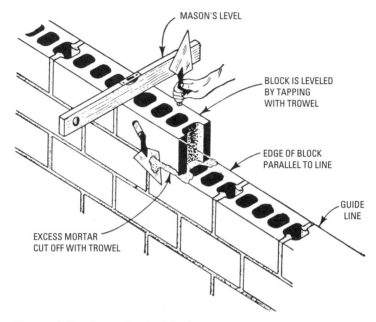

MASON'S LEVEL

BLOCK IS LEVELED
BY TAPPING
WITH TROWEL

EDGE OF BLOCK
PARALLEL TO LINE

GUIDE
LINE

EXCESS MORTAR
CUT OFF WITH TROWEL

Figure 5-7   A method of laying concrete blocks. Good workmanship requires straight courses with face of wall plumb.

is customary to lay up the corner blocks three or four courses high and use them as guides in laying the rest of the walls.

A full width of mortar is placed on the footing (see Figure 5-9), with the first course laid two or three blocks long each way from the corner. The second course is a half-block shorter each way than the first course; the third course is a half-block shorter than the second, and so on. Thus, the corners are stepped back until only the corner blocks are laid. Use a line and level frequently to see that the blocks are laid straight and that the corners are plumb. It is customary that such special units as corner blocks, door- and window-jamb blocks, fillers, and veneer blocks be provided prior to starting the laying of the blocks.

Building Walls Between Corners

In laying walls between corners, a line is stretched tightly from corner to corner to serve as a guide (see Figure 5-10). The line is fastened to nails or wedges driven into the mortar joints, so that when stretched it just touches the upper outer edges of the block laid in

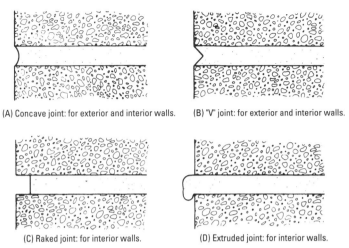

(A) Concave joint: for exterior and interior walls.

(B) "V" joint: for exterior and interior walls.

(C) Raked joint: for interior walls.

(D) Extruded joint: for interior walls.

Figure 5-8   Four joint styles popular in block wall construction.

the corners. The blocks in the wall between corners are laid so that they will just touch the cord in the same manner. In this way, straight horizontal joints are secured. Prior to laying up the outside wall, the door and window frames should be on hand to set in place as guides for obtaining the correct opening.

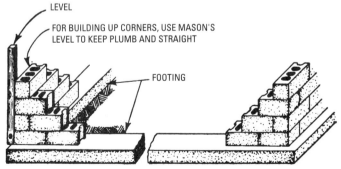

LEVEL

FOR BUILDING UP CORNERS, USE MASON'S LEVEL TO KEEP PLUMB AND STRAIGHT

FOOTING

Figure 5-9   Laying up corners when building with concrete masonry-block units.

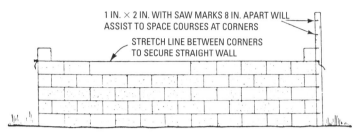

Figure 5-10 Procedure for laying concrete-block walls.

Building around Door and Window Frames

There are several acceptable methods of building door and window frames in concrete masonry walls (see Figure 5-11). One method used is to set the frames in the proper position in the wall. The frames are then plumbed and carefully braced, after which the walls are built up against them on both sides. Concrete sills may be poured later.

The frames are often fastened to the walls by anchor bolts passed through the frames and embedded in the mortar joints. Another method of building frames in concrete masonry walls is to build openings for them by using special jamb blocks (see Figure 5-12). The frames are inserted after the wall is built. The only advantage to this method is that the frames can be taken out without damaging the wall, should it ever become necessary.

Placing Sills and Lintels

Building codes require that concrete-block walls above openings be supported by arches or lintels of metal or masonry (plain or reinforced). The arches or lintels must extend into the walls not less than 4 inches on each side. Stone or other nonreinforced masonry lintels should not be used unless supplemented on the inside of the wall with iron or steel lintels (see Figure 5-13).

These are usually prefabricated, but may be made up on the job if desired. Lintels are reinforced with steel bars placed 1½ inch from the lower side. The number and size of reinforcing rods depend upon the width of the opening and the weight of the load to be carried.

Sills serve the purpose of providing watertight bases at the bottom of wall openings. Since they are made in one piece, there are no joints for possible leakage of water into the walls below. They are sloped on the top face to drain water away quickly. They are usually made to project 1½ to 2 inches beyond the wall face, and are made with a groove along the lower outer edge to provide a drain so that water

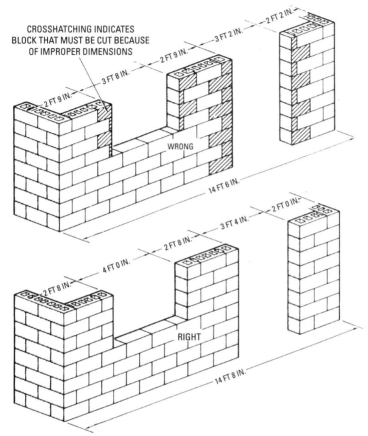

CROSSHATCHING INDICATES
BLOCK THAT MUST BE CUT BECAUSE
OF IMPROPER DIMENSIONS

Figure 5-11 **The right and wrong ways to plan door and window openings in block walls.** *(Courtesy Portland Cement Association)*

dripping off the sill will fall free and not flow over the face of the wall (possibly causing staining).

Slip sills are popular because they can be inserted after the wall has been built. Therefore, they require no protection during construction. Since there is an exposed joint at each end of the sill, special care should be taken to see that it is completely filled with mortar and the joints packed tight.

Lug sills are projected into the concrete block wall (usually 4 inches at each end). The projecting parts are called *lugs*. There

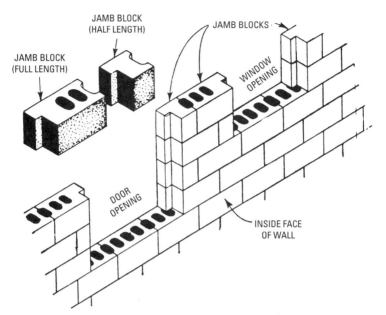

Figure 5-12 A method of laying openings for doors and windows.

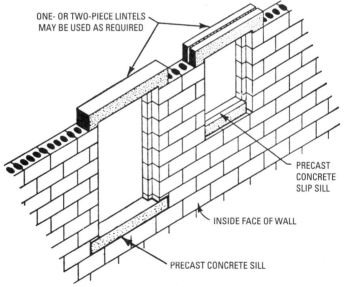

Figure 5-13 A method of inserting precast concrete lintels and sills in concrete block walls.

are no vertical mortar joints at the juncture of the sills and the jambs. Like slip sills, lug sills are usually made to project from $1\frac{1}{2}$ to 2 inches over the face of the wall. The sill is provided with a groove under the lower outer edge to form a drain. Frequently they are made with washes at either end to divert water away from the juncture of the sills and the jambs. This is in addition to the outward slope on the sills.

At the time the lug sills are set, only the portion projecting into the wall is bedded in mortar. The portion immediately below the wall opening is left open and free of contact with the wall below. This is done in case there is minor settlement or adjustments in the masonry work during construction, thus avoiding possible damage to the sill during the construction period.

Construction Methods

The steps in building concrete masonry walls are:

1. Mixing the mortar
2. Building the wall between corners
3. Building around door and window frames
4. Placing sills and lintels
5. Building interior walls
6. Attaching sills and plates

Walls

Thickness of concrete masonry walls is usually governed by building codes, if any are in existence at the particular location. Eight inches is generally specified as the minimum thickness for all exterior walls, and for load-bearing interior walls. Partitions and curtain walls are often made 3, 4, or 6 inches thick. The thickness of bearing walls in heavily loaded buildings is properly governed by the loading.

Garden walls less than 4 feet high can be as thin as 4 inches, but it is best to make them 8 inches thick. Walls over 4 feet high must be at least 8 inches thick to provide sufficient strength.

A wall no more than 4 feet high needs no reinforcement. Merely build up the fence from the foundation with block or brick and mortar. Over 4 feet, however, reinforcement will be required, and the fence should be of block, not brick. Set $\frac{1}{2}$-inch-diameter steel rods in the poured concrete foundation at 4-foot centers (see Figure 5-2). When you have laid the blocks (with mortar) up to the level of the top of the rods, pour concrete into the hollow cores around the rods. Then continue on up with the rest of the layers of blocks.

In areas subject to possible earthquake shocks or extra-high winds, horizontal reinforcement bars should also be used in high walls. Use No. 2 (¹/₄-inch) bars or special straps made for the purpose. Figure 5-14 is a photograph of a block wall based on Figure 5-2. The foundation is concrete poured in a trench dug out of the ground. Horizontal reinforcement is in the concrete, with vertical members bent up at intervals. High column blocks (16 inches × 16 inches × 8 inches) are laid at the vertical rods. The columns are evident in the finished wall shown in Figure 5-15.

Load-bearing walls are those used as exterior and interior walls in residential and industrial buildings. Not only must the wall support the roof structure, but also it must bear its own weight. The greater the number of stories in the building, the greater the thickness the

Figure 5-14   Vertical reinforcement rods through double-thick column blocks.

Figure 5-15   A newly finished concrete-block wall. Note reinforcement columns at regular intervals.

lower stories must be to support the weight of the concrete blocks above, as well as the roof structure.

Basement walls should not be narrower than the walls immediately above them, and not less than 12 inches for unit masonry walls. Solid cast-in-place concrete walls are reinforced with at least one $^3/_8$-inch deformed bar (spaced every 2 feet) continuous from the footing to the top of the foundation wall. Basement walls with 8-inch hollow concrete blocks frequently prove very troublesome. All hollow-block foundation walls should be capped with a 4-inch solid concrete block, or the core should be filled with concrete.

Building Interior Walls

Interior walls are built in the same manner as exterior walls. Load-bearing interior walls are usually made 8 inches thick. Partitions that are not load-bearing walls are usually 4 inches thick. Figure 5-16 shows a method of joining interior load-bearing walls to exterior walls.

Sills and Plates

Sills and plates are usually attached to concrete block walls by means of anchor bolts (see Figure 5-17). These bolts are placed in the cores of the blocks, and the cores are filled with concrete. The bolts are spaced about 4 feet apart under average conditions. Usually, $^1/_2$-inch bolts are used and should be long enough to go through two courses of blocks and project through the plate about an inch to permit the use of a large washer and the anchor-bolt nut.

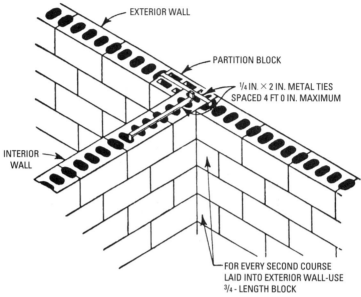

Figure 5-16 Showing detail of joining interior and exterior wall in concrete block construction.

Installation of Heating and Ventilating Ducts

These ducts are provided for as shown on the architect's plans. The placement of the heating ducts depends upon the type of wall, whether it is load-bearing or not. Figure 5-18 shows a typical example of placing the heating or ventilating ducts in an interior concrete masonry wall.

Interior concrete-block walls that are not load-bearing, and that are to be plastered on both sides, are frequently cut through to provide for the heating duct, the wall being flush with the ducts on either side. Metal lath is used over the ducts.

Electrical Outlets

These outlets are provided for by inserting outlet boxes in the walls (see Figure 5-19). All wiring should be installed to conform to the requirements of the *National Electrical Code* or local codes.

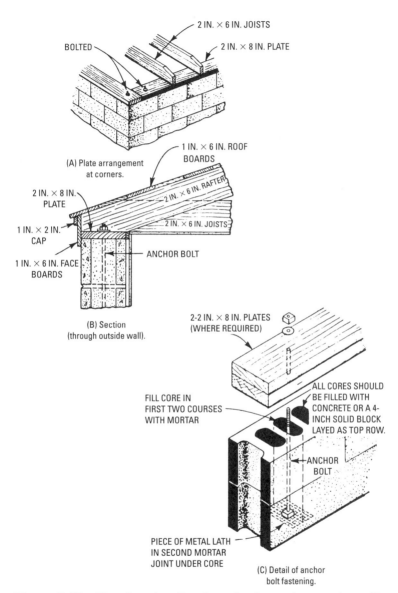

2 IN. × 6 IN. JOISTS

2 IN. × 8 IN. PLATE

BOLTED

(A) Plate arrangement at corners.

1 IN. × 6 IN. ROOF BOARDS

2 IN. × 6 IN. RAFTER

2 IN. × 6 IN. JOISTS

2 IN. × 8 IN. PLATE

1 IN. × 2 IN. CAP

1 IN. × 6 IN. FACE BOARDS

ANCHOR BOLT

(B) Section (through outside wall).

2-2 IN. × 8 IN. PLATES (WHERE REQUIRED)

FILL CORE IN FIRST TWO COURSES WITH MORTAR

ALL CORES SHOULD BE FILLED WITH CONCRETE OR A 4-INCH SOLID BLOCK LAYED AS TOP ROW.

ANCHOR BOLT

PIECE OF METAL LATH IN SECOND MORTAR JOINT UNDER CORE

(C) Detail of anchor bolt fastening.

Figure 5-17 Showing details of methods used to anchor sills and plates to concrete block walls.

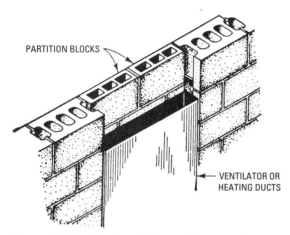

PARTITION BLOCKS

VENTILATOR OR
HEATING DUCTS

Figure 5-18   Method of installing ventilating and heating duct in concrete block walls.

Insulation

Block walls can (and should) be insulated in a number of ways. One way is to secure studs to the walls with cut nails or other fasteners, then staple batt insulation to the studs. However, a simpler way is with Styrofoam insulation. This comes in easy-to-cut sheets. It is applied to the interior walls with mastic and then, because it is

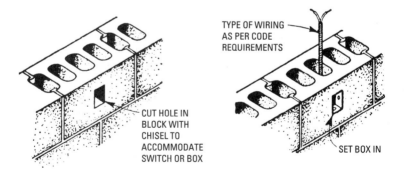

TYPE OF WIRING
AS PER CODE
REQUIREMENTS

CUT HOLE IN
BLOCK WITH
CHISEL TO
ACCOMMODATE
SWITCH OR BOX

SET BOX IN

Figure 5-19   Method showing installation of electrical switches and outlet boxes in concrete-block walls.

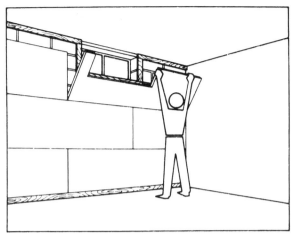

Figure 5-20   Styrofoam rigid insulation board can be glued to masonry.

flammable, covered with gypsum drywall or other noncombustible material (see Figure 5-20).

The Styrofoam may also be used on the exterior of the wall. In this case, it is secured only to that part of the foundation above the earth and one foot below. The Styrofoam, in turn, is covered with cement, or other mixture as specified by the manufacturer.

Another way is by filling the cores of concrete block units in all outside walls with granulated insulation or, the preferred way, by inserting rigid foam insulation.

Flashing
Adequate flashing with rust- and corrosion-resisting material is of the utmost importance in masonry construction, since it prevents water from getting through the walls at vulnerable points. Points requiring protection by flashing are:

- Tops and sides of projecting trim
- Under coping and sills
- At intersection of wall and roof
- Under built-in gutters
- At intersection of chimney and roof
- At all other points where moisture is likely to gain entrance

Flashing material usually consists of No. 26-gage (14-ounce) copper sheet or other approved noncorrodible material.

Floors

In concrete masonry construction, floors may be made entirely of wood or concrete, although a combination of concrete and wood is sometimes used. Wooden floors (see Figure 5-21) must be framed with one governing consideration: they must be made strong enough to carry the load. The type of building and its use will determine the type of floor used, the thickness of the sheathing, and approximate spacing of the joists. The girder is usually made of heavy

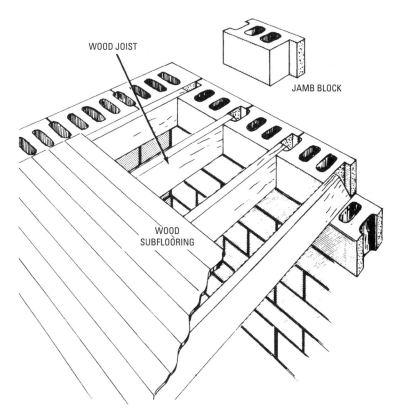

WOOD JOIST

JAMB BLOCK

WOOD SUBFLOORING

Figure 5-21 Method of installing wooden floor joists in concrete blocks.

timber and is used to support the lighter joists between the outside walls.

Concrete Floors
There are usually two types of concrete floors:

- Cast-in-place concrete
- Precast joists used with cast-in-place or precast concrete units

Cast-in-Place
Cast-in-place concrete floors (Figure 5-22), are used in basements (and in residences without basements) and are usually reinforced by means of wire mesh to provide additional strength and to prevent cracks in the floor. This type of floor is used for small areas. It used to be popular for porch floors.

Figure 5-23 shows another type of cast-in-place concrete floor. This floor has more strength because of the built-in joist. It can be used for larger areas.

Precast Joists
Figure 5-24 shows the precast-joist type. The joists are set in place, the wooden forms inserted, the concrete floor poured, and the forms

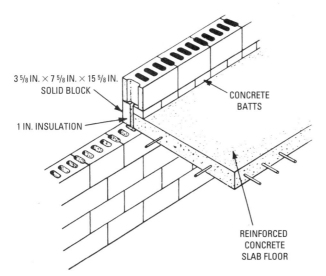

3 ⅝ IN. × 7 ⅝ IN. × 15 ⅝ IN.
SOLID BLOCK

1 IN. INSULATION

CONCRETE BATTS

REINFORCED CONCRETE SLAB FLOOR

Figure 5-22 Cast-in-place concrete floor supported by the concrete block wall.

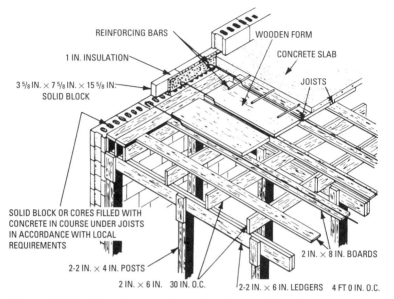

REINFORCING BARS

WOODEN FORM

1 IN. INSULATION

CONCRETE SLAB

3 5/8 IN. × 7 5/8 IN. × 15 5/8 IN. SOLID BLOCK

JOISTS

SOLID BLOCK OR CORES FILLED WITH CONCRETE IN COURSE UNDER JOISTS IN ACCORDANCE WITH LOCAL REQUIREMENTS

2 IN. × 8 IN. BOARDS

2-2 IN. × 4 IN. POSTS

2 IN. × 6 IN.   30 IN. O.C.

2-2 IN. × 6 IN. LEDGERS   4 FT 0 IN. O.C.

Figure 5-23   Framing construction for cast-in-place concrete floor.

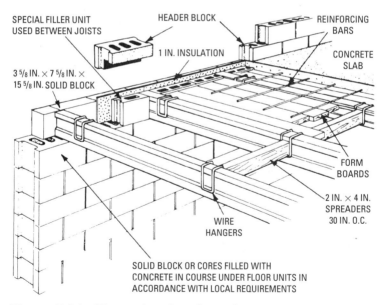

SPECIAL FILLER UNIT USED BETWEEN JOISTS

HEADER BLOCK

REINFORCING BARS

1 IN. INSULATION

CONCRETE SLAB

3 5/8 IN. × 7 5/8 IN. × 15 5/8 IN. SOLID BLOCK

FORM BOARDS

2 IN. × 4 IN. SPREADERS 30 IN. O.C.

WIRE HANGERS

SOLID BLOCK OR CORES FILLED WITH CONCRETE IN COURSE UNDER FLOOR UNITS IN ACCORDANCE WITH LOCAL REQUIREMENTS

Figure 5-24   Illustrating framing of precast concrete-floor joists.

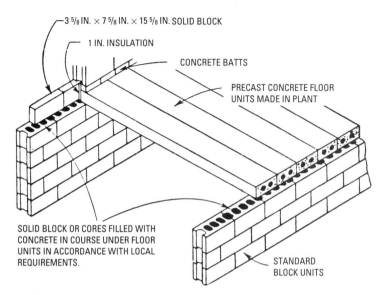

3 5/8 IN. × 7 5/8 IN. × 15 5/8 IN. SOLID BLOCK

1 IN. INSULATION

CONCRETE BATTS

PRECAST CONCRETE FLOOR
UNITS MADE IN PLANT

SOLID BLOCK OR CORES FILLED WITH
CONCRETE IN COURSE UNDER FLOOR
UNITS IN ACCORDANCE WITH LOCAL
REQUIREMENTS.

STANDARD
BLOCK UNITS

Figure 5-25   Method of framing cored concrete-floor units into concrete walls.

removed. Another type of precast-joist concrete floor (see Figure 5-25) consists of precast concrete joists covered with cast-in-place concrete slabs. The joists are usually made in a concrete-products plant.

Usual spacing of the joists is from 27 to 33 inches, depending upon the span and the load. The cast-in-place concrete slab is usually 2 or $2^{1}/_{2}$ inches thick, and extends down over the heads of the joists about $^{1}/_{2}$ inch. Precast concrete joists are usually made in 8-, 10-, and 12-inch depths. The 8-inch joists are used for spans up to 16 feet; the 10-inch joists are for spans between 16 and 20 feet; and the 12-inch joists are for spans from 20 to 24 feet.

It is customary to double the joists under the partition where masonry partitions do not support a load. Moreover, they are placed parallel to the joists. If the partition runs at right angles to the joists, the usual practice is to design the floor to carry an additional load of 20 pounds per square foot. Precast concrete joists may be left exposed on the underside and painted, or a suspended ceiling may be used. An attractive variation in exposed joist treatment is to double the joists and increase the spacing. Where this is done, the concrete slab is made $2^{1}/_{2}$ inches thick for spacings up to 48 inches, and 3 inches thick for spacings from 48 to 60 inches. Joists may be doubled by setting them close together, or by leaving a space between them and filling the space with concrete.

Wood Floors

Where wood-surfaced floors are used in homes, any type of hardwood (such as maple, birch, beech, or oak) may be laid over the structural concrete floor. The standard method of laying hardwood flooring over a structural-concrete floor slab is to nail the boards to 2-inch × 2-inch or 2-inch × 3-inch sleepers. These sleepers may be tied to the tops of the stirrups protruding from the precast concrete joists. They should be placed not more than 16 inches on center, preferably 12 inches in residential construction. Before the hardwood flooring is nailed to the sleepers, the concrete floor must be thoroughly hardened and free from moisture. It is also best to delay laying the hardwood floors until after drywall installers or plasterers have finished work.

Certain types of parquet and design wood floorings are laid directly on the concrete, being bonded to the surface with adhesive. The concrete surface should be troweled smooth and be free from moisture. No special topping is required. The ordinary level-sidewalk finish provides a satisfactory base. Manufacturer's directions for laying this type of wood flooring should be followed carefully to ensure satisfactory results in the finished floor.

Summary

Concrete blocks provide a very good building material. Blocks are very reasonable in cost, durable, fire-resistant, and can carry heavy loads. The thickness of concrete blocks is generally 8 inches (except for partition blocks), which is generally specified as the minimum thickness for all exterior walls and for load-bearing interior walls. Partitions and curtain walls are often made 3, 4, or 6 inches thick.

Concrete blocks are made in several sizes and shapes. For various openings (such as windows or doors), a jamb block is generally used to adapt to various frame designs. Building codes require that concrete block walls above openings shall be supported by arches of steel or precast masonry lintels. Lintels are usually prefabricated, but may be made on the job if desired. They are reinforced with steel bars placed approximately $1\frac{1}{2}$ inches from the lower side.

In concrete-block construction, floors may be entirely of wood or concrete, although a combination of the two materials is sometimes used. Joists may be installed in the block wall by using jamb blocks, or by installing a sill plate at the top of the wall. Concrete floors are generally of the cast-in-place, precast unit, or precast-joist types.

Review Questions

1. What is the general thickness of concrete blocks?
2. What are the standard measurements of a full-size concrete block?
3. What are lintels? Why are they used?
4. What are precast joists?
5. What are concrete bats?
6. How is a jamb block different from a regular block?
7. What are the two types of concrete floors?
8. How is a partition block different from regular building blocks?
9. How far do lug sills project into the concrete block?
10. Why are slip sills popular?

Chapter 6

Frames and Framing

Good material and workmanship will be of very little value unless the underlying framework of a building is strong and rigid. The resistance of a house to such forces as tornadoes and earthquakes, and control of cracks caused by settlement, all depends on a good framework.

Methods of Framing

Although it is true that no two buildings are put together in exactly the same manner, disagreement exists among architects and carpenters as to which method of framing will prove most satisfactory for a given condition. *Light-framed construction* may be classified into the following three distinct types:

- Balloon frame
- Post-and-beam
- Platform frame

Balloon-Frame Construction

The principal characteristic of *balloon framing* is the use of studs extending in one piece from the foundation to the roof (see Figure 6-1). The joists are nailed to the studs and supported by a ledger board set into the studs. Diagonal sheathing may be used instead of wallboard to eliminate corner bracing.

Post-and-Beam Construction

The *post-and-beam construction* is the oldest method of framing used in the United States. It is still used, particularly where the builder wants large open areas in a house. This type of framing is characterized by heavy timber posts (see Figure 6-2), often with intermediate posts between.

Platform Frame Construction

Platform (or *western*) *framing* is characterized by platforms independently framed, the second or third floor being supported by the studs from the first floor (see Figure 6-3). The chief advantage in this type of framing (in all-lumber construction) lies in the fact that if there is any settlement caused by shrinkage, it will be uniform throughout and will not be noticeable.

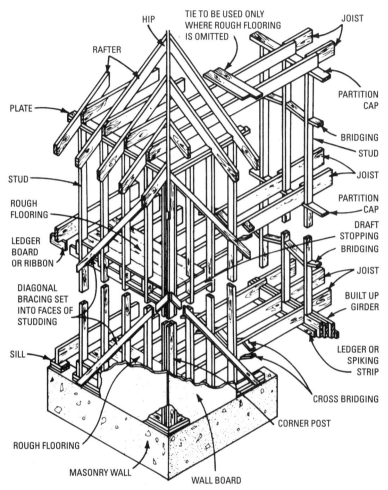

Figure 6-1   Details of balloon-frame construction.

Framing Terms

The following sections describe terms and concepts commonly used in reference to framing.

Sills

The *sills* are the first part of the framing to be set in place. Sills may rest directly on foundation piers or other type of foundation, and

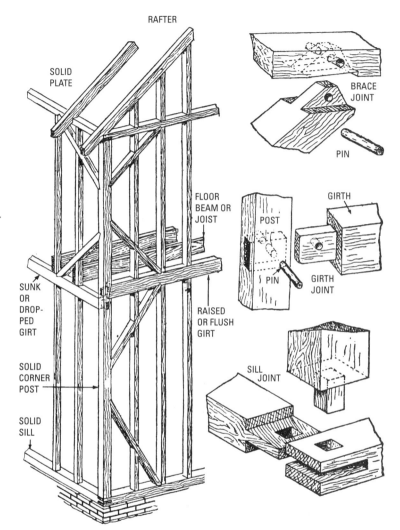

Figure 6-2 Post-and-beam frame construction. Heavy solid timbers are fastened together with pegged mortise-and-tenon joints.

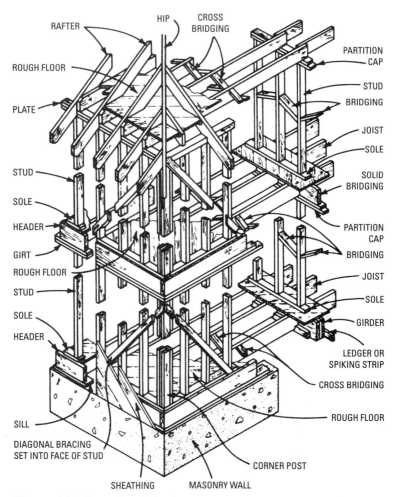

Figure 6-3 Details of platform, or western, construction.

usually extend all around the building. When *box sills* are used, the part of the sill that rests on the foundation wall is called the *sill plate*.

Girders
A *girder* may be a single beam or built-up thinner lumber (see Figure 6-4). Girders usually support joists, whereas the girders themselves

are supported by bearing walls or columns. When a girder is supported by a wall or column, it must be remembered that such a member delivers a large concentrated load to a small section of the wall or column. Therefore, care must be taken to see that the wall or column is strong enough to carry the load imposed upon it by the girder. Girders are generally used only where the joist will not safely span the distance. The size of a girder is determined by the span length and the load to be carried.

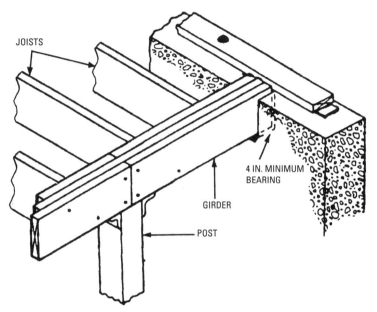

Figure 6-4   Built-up wooden girders.

Joists
Joists are the pieces that make up the body of the floor frame, and to which the flooring and subflooring are nailed (see Figure 6-5). They are usually 2 or 3 inches thick, with depth varying to suit conditions. Joists carry a *dead load* (composed of the weight of the joists themselves, in addition to the flooring) and a *live load* (composed of the weight of furniture and persons).

Subflooring
A *subfloor* is laid on the joists and nailed to them (see Figure 6-5). By the use of subflooring, floors are made much stronger, since the

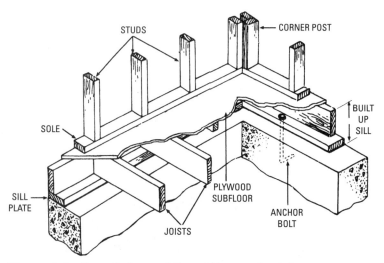

Figure 6-5 Detail views of floor joists and subfloor.

floor weight is distributed over a larger area. It may be laid before or after the walls are framed, but preferably before. The subfloor can then be used as a floor to work on while framing the walls. The material for subflooring can be sheathing boards (see Figure 6-6) or plywood (see Figure 6-5).

Headers and Trimmers

When an opening is to be made in a floor, a *header* or *trimmer* holds the ends of the joists. The header or trimmer should be made heavier than the joists to carry the extra load. The header must be framed in between the trimmers to receive the ends of the short joists.

Walls and Partitions

All *walls* and *partitions* in which the structural elements are wood are classed as frame construction. Their structural elements are closely spaced. They contain a number of vertical members called studs (see Figure 6-5). These are arranged in a row with their ends bearing on a long longitudinal member called the bottom plate or *sole plate*, and their tops are capped with another plate called the *top plate*. The bearing strength of the stud walls is governed by the length of the studs.

The top plate serves two purposes: to tie the studding together at the top and form a finish for the walls, and to furnish a support

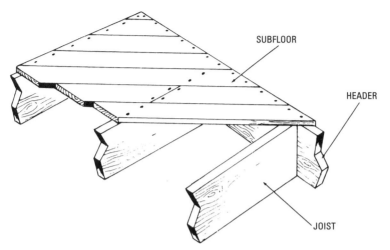

Figure 6-6   Method of laying lumber subflooring.

for the lower end of the rafters. The top plate further serves as a connecting link between the ceiling and the walls.

The plate is made up of one or two pieces of timber of the same size as the studs. In cases where studs at the end of the building extend to the rafters, no plate is used. When it is used on top of partition walls, it is sometimes called a *cap*. The sole plate is the bottom horizontal member on which the studs rest.

Partition walls are any walls that divide the inside space of a building. In most cases, these walls are framed as a part of the building. In cases where floors are to be installed after the outside of the building is completed, the partition walls are left unframed. There are two types of partition walls, bearing and non-bearing. The *bearing* type (also called *load-bearing*) supports the structure above. The *nonbearing* type supports only itself. It may be put in at any time after the framework is installed. Only one cap or plate is used.

A sole plate should be used in every case, as it helps to distribute the load over a large area. Partition walls are framed in the same manner as outside walls, and inside door openings are framed the same as outside openings. Where there are corners (or where one partition wall joins another), corner posts or T-posts are used in the outside walls. These posts provide nailing surfaces for the inside wall finish.

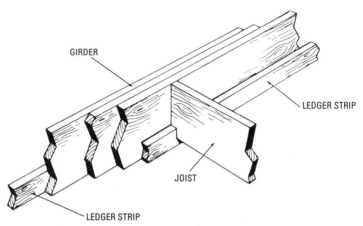

Figure 6-7   Detail of girder and method of supporting joists by using a ledger strip.

Ledger Plates

In connecting joists to girders and sills where piers are used, a 2-inch × 4-inch piece is nailed to the face of the sill or girder and flush with the bottom edge, as shown in Figure 6-7. This is called a *ledger*.

Braces

Braces are used as a permanent part of the structure. They stiffen the walls and keep corners square and plumb. Braces also prevent the frame from being distorted by lateral forces (such as by wind or by settlement). These braces are placed wherever the sills or plates make an angle with the corner post or with a T-post in the outside wall. The brace extends from the sill or sole plate to the top of the post, forming an angle of approximately 60° with the sole plate and an angle of 30° with the post.

Studs

Studs are the main vertical framing members. They are installed between the top and sole plates. Before studs are laid out, the windows and door openings are laid out. The studs are normally set 16 inches apart (on centers), but there is a trend today toward 24-inch centers. In most instances, the studs are nailed to the top and sill plates while the entire assembly is resting flat on the deck. The wall is then raised into position.

Bridging

Walls of frame buildings are bridged in most cases in order to increase their strength. There are two methods of *bridging*: the diagonal and the horizontal. *Diagonal bridging* is nailed between the studs at an angle. It is more effective than horizontal bridging, because it forms a continuous truss and tends to keep the walls from sagging. Whenever possible, both outside and inside walls should be bridged alike.

Horizontal bridging is nailed between the studs, halfway between the sole and top plate. This bridging is cut to lengths that correspond to the distance between the studs at the bottom. Such bridging not only stiffens the wall, but also helps straighten the studs.

Metal bridging is made of forged thin strips of metal secured to tops of joists.

Rafters

In all roofs, the pieces that make up the main body of the framework are called the *rafters*. They do for the roof what the joists do for the floor or the studs do for the wall. The rafters are inclined members, usually spaced from 16 to 24 inches on center, that rest at the bottom on the plate and are fastened on center at the top in various ways according to the form of the roof. The plate forms the connecting link between the wall and the roof and is really a part of both.

The size of the rafters varies, depending upon the length and the distance at which they are spaced. The connection between the rafter and the wall is the same in all types of roofs. They may or may not extend out a short distance from the wall to form the eaves and to protect the sides of the building.

Lumber Terms

The basic construction material in carpentry is lumber. There are many kinds of lumber, and they vary greatly in structural characteristics. Here, we deal with lumber common to construction carpentry, its application, the standard sizes in which it is available, and the methods of computing lumber quantities in terms of board feet.

Standard Sizes of Lumber

Lumber is usually sawed into standard lengths, widths, and thicknesses. This permits uniformity in planning structures and in ordering material. Table 6-1 lists the common widths and thicknesses of wood in rough and in dressed dimensions in the United States. Standards have been established for dimension differences between *nominal size* and the *standard size* (which is actually the reduced size

Table 6-1   Guide to Size of Lumber

What You Order	What You Get	
	Dry or Seasoned*	Green or Unseasoned**
1×4	$^3/_4 \times 3^1/_2$	$^{25}/_{32} \times 3^9/_{16}$
1×6	$^3/_4 \times 5^1/_2$	$^{25}/_{32} \times 5^5/_8$
1×8	$^3/_4 \times 7^1/_4$	$^{25}/_{32} \times 7^1/_2$
1×10	$^3/_4 \times 9^1/_4$	$^{25}/_{32} \times 9^1/_2$
1×12	$^3/_4 \times 11^1/_4$	$^{25}/_{32} \times 11^1/_2$
2×4	$1^1/_2 \times 3^1/_2$	$1^9/_{16} \times 3^9/_{16}$
2×6	$1^1/_2 \times 5^1/_2$	$1^9/_{16} \times 5^5/_8$
2×8	$1^1/_2 \times 7^1/_4$	$1^9/_{16} \times 7^1/_2$
2×10	$1^1/_2 \times 9^1/_4$	$1^9/_{16} \times 9^1/_2$
2×12	$1^1/_2 \times 11^1/_4$	$1^9/_{16} \times 11^1/_2$
4×4	$3^1/_2 \times 3^1/_2$	$3^9/_{16} \times 3^9/_{16}$
4×6	$3^1/_2 \times 5^1/_2$	$3^9/_{16} \times 5^5/_8$
4×8	$3^1/_2 \times 7^1/_4$	$3^9/_{16} \times 7^1/_2$
4×10	$3^1/_2 \times 9^1/_4$	$3^9/_{16} \times 9^1/_2$
4×12	$3^1/_2 \times 11^1/_4$	$3^9/_{16} \times 11^1/_2$

*19 percent moisture content or less.
**More than 19 percent moisture content.

when dressed). It is important that these dimension differences be taken into consideration when planning a structure. A good example of the dimension differences may be illustrated by the common 2×4. As may be seen in the table, the familiar quoted size (2×4) refers to a rough or nominal dimension, but the actual standard size to which the lumber is dressed is $1^1/_2 \times 3^1/_2$.

Framing Lumber

The frame of a building consists of the wooden form constructed to support the finished members of the structure. It includes such items as posts, girders (beams), joists, subfloor, sole plate, studs, and rafters. Softwoods are usually used for lightwood framing and all other aspects of construction carpentry considered in this book. They are cut into the standard sizes required for light framing (including 2×4, 2×6, 2×8, 2×10, 2×12, and all other sizes required for framework), with the exception of those sizes classed as structural lumber.

Although No. 1 and No. 3 common are sometimes used for framing, No. 2 common is most often used, and is, therefore, most often stocked and available in retail lumberyards in the common sizes for various framing members. However, the size of lumber required for any specific structure will vary with the design of the building (such as light-frame or heavy-frame,) and the design of the particular members (such as beams or girders).

The exterior walls of a frame building usually consist of three layers: sheathing, building paper, and siding. Sheathing lumber is usually 1 × 6 boards or 1 × 8 boards, No. 2 or No. 3 common softwood. It may be plain, tongue-and-groove, or ship lapped. The siding lumber may be grade C, which it most often used. Siding is usually procured in bundles consisting of a given number of square feet per bundle, and comes in various lengths up to a maximum of 20 feet. Sheathing grade (CDX) plywood is used more often than lumber for sheathing.

Summary

After the foundation walls have been completed, the first part of framing is to set the sills in place. The sills usually extend all around the building. Where double or built-up sills are used, the joints are staggered and the corner joints are lapped.

Where floor joists span a large distance, girders are used to help support the load. The size of a girder depends on the span length of the joists. Girders may be either a steel I-beam, or can be built on the job from 2 × 8 or 2 × 10 lumber placed side-by-side.

Joists are the pieces that make up the body of the floor frame, and to which the subflooring is nailed. The joists are usually 2 × 6 or 2 × 8 lumber commonly spaced 16 or 24 inches apart. The subflooring is generally installed before the wall construction is started, and in this manner, the subfloor can serve as a work floor.

Review Questions

1. Explain the purpose of the sill plate, floor joists, and girder.
2. What is the purpose of the subflooring?
3. What is a ledger strip?
4. What is a sole plate?
5. Where are headers and trimmers used?
6. How can light-framed construction be classified?
7. What is meant by balloon framing?
8. How is post-and-beam type construction anchored to the foundation?
9. What is a joist?
10. What is a cap?

Chapter 7

Floors, Girders, and Sills

By definition, a *girder* is a principal beam extending from wall-to-wall of a building, affording support for the joists or floor beams where the distance is too great for a single span. Girders may be either built-up wood (see Figure 7-1) or steel (see Figure 7-2).

Girders

There are girders that have been engineered to fit a particular job. Shopping malls, churches, and schools all require long girders to support an open space for a variety of reasons. The construction of girders and the placement of them are important to carpenters and designers as well as builders.

Construction of Girders

Girders can be built up of wood if select stock is used. Be sure it is straight and sound (see Figure 7-1). If the girders are to be built up of 2 × 8 or 2 × 10 stock, place the pieces on the sawhorses and nail them together. Use the piece of stock that has the least amount of warp for the centerpiece and nail other pieces on the sides of the center stock. Use a common nail that will go through the first piece and nearly through the centerpiece. Square off the ends of the girder after the pieces have been nailed together. If the stock is not long enough to build up the girder the entire length, the pieces must be built up by staggering the joints. If the girder supporting post is to be built up, it is to be done in the same manner as described for the girder.

Table 7-1 is an example of sizes of built-up wood girders for various loads and spans, based on Douglas fir 4-square framing lumber. All girders are figured as being made with 2-inch dressed stock. The 6-inch girder is figured three pieces thick; the 8-inch girder four pieces thick; the 10-inch girder five pieces thick, and the 12-inch girder six pieces thick. For solid girders, multiply the load in Table 7-1 by 1.130 when 6-inch girders are used; 1.150 when 8-inch girders are used; 1.170 when 10-inch girders are used, and 1.180 when 12-inch girders are used. Other woods will require figures based on the specific wood and grade.

Placing Basement Girders

Basement girders must be lifted into place on top of the piers, and walls built for them, and set perfectly level and straight from end

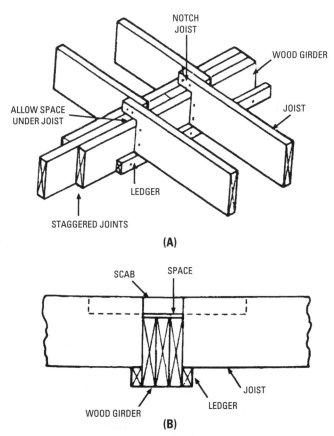

NOTCH
JOIST

WOOD GIRDER

ALLOW SPACE
UNDER JOIST

JOIST

LEDGER

STAGGERED JOINTS

(A)

SCAB   SPACE

JOIST

WOOD GIRDER   LEDGER

(B)

Figure 7-1   Built-up wood girder: (A) the floor joists are notched to fit over the girder; (B) a connecting scab is used to tie joists together.

to end. Some carpenters prefer to give the girders a slight crown of approximately 1 inch in the entire length, which is a wise plan because the piers will generally settle. They settle a little more than the outside walls.

When there are posts instead of brick piers used to support the girder, the best method is to temporarily support the girder by uprights made of 2 × 4 joists resting on blocks on the ground below.

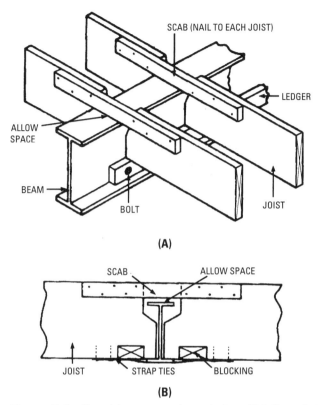

SCAB (NAIL TO EACH JOIST)

LEDGER

ALLOW SPACE

BEAM

BOLT

JOIST

(A)

SCAB   ALLOW SPACE

JOIST   STRAP TIES   BLOCKING

(B)

Figure 7-2   Steel beam used as girder: (A) floor joists rest on wooden ledger bolted to steel I-beam girder; (B) floor joists rest directly on bottom flange of I-beam. The joists are connected above with a scab board.

When the superstructure is raised, these can be knocked out after the permanent posts are placed. The practice of temporarily shoring the girders, and not placing the permanent supports until after the superstructure is finished, is favored by many builders, and it is a good idea for carpenters to know just how it should be done. Permanent supports are usually made by using 3- or 4-inch steel pipe set in a concrete foundation, or footing that is at least 12 inches deep. They are called *Lally columns* and are filled with concrete for greater strength.

Table 7-1 Examples of Girder Sizes

Load per Linear Foot of Girder	Length of Span				
	6-foot	*7-foot*	*8-foot*	*9-foot*	*10-foot*
750	6 × 8 in.	6 × 8 in.	6 × 8 in.	6 × 10 in.	6 × 10 in.
900	6 × 8	6 × 8	6 × 10	6 × 10	8 × 10
1050	6 × 8	6 × 10	8 × 10	8 × 10	8 × 12
1200	6 × 10	8 × 10	8 × 10	8 × 10	8 × 12
1350	6 × 10	8 × 10	8 × 10	8 × 12	10 × 12
1500	8 × 10	8 × 10	8 × 12	10 × 12	10 × 12
1650	8 × 10	8 × 12	10 × 12	10 × 12	10 × 14
1800	8 × 10	8 × 12	10 × 12	10 × 12	10 × 14
1950	8 × 12	10 × 12	10 × 12	10 × 14	12 × 14
2100	8 × 12	10 × 12	10 × 14	12 × 14	12 × 14
2250	10 × 12	10 × 12	10 × 14	12 × 14	12 × 14
2400	10 × 12	10 × 14	10 × 14	12 × 14	
2550	10 × 12	10 × 14	12 × 14	12 × 14	
2700	10 × 12	10 × 14	12 × 14		
2850	10 × 14	12 × 14	12 × 14		
3000	10 × 14	12 × 14			
3150	10 × 14	12 × 14			
3300	12 × 14	12 × 14			

Sills

A *sill* is that part of the sidewalls of a house that rests horizontally upon, and is securely fastened to, the foundation. There are various types of sills and they may be divided into two general classes:

- Solid
- Built-up

The *built-up sill* has become more or less a necessity because of the high cost and scarcity of larger-sized timber. The work involved in sill construction is very important for the carpenter. The foundation wall is the support upon which the entire structure rests. The sill is the foundation on which all the framing structure rests, and is the real point of departure for actual carpentry and joinery activities. The sills are the first part of the frame to be set in place. They rest directly on the foundation piers or on the ground, and may extend all around the building.

Types of Sills
The type of sill used depends upon the general type of construction. Types of sills include the following:

- Box sills
- T-sills
- Braced-framing sills
- Built-up sills

Box Sills
Box sills are often used with the very common style of platform framing, either with or without the sill plate (see Figure 7-3). In this type of sill, the part that lies on the foundation wall or ground is called the *sill plate*. The sill is laid edgewise on the outside edge of the sill plate. It is doubled and bolted to the foundation. Aluminum is set in asphaltum beneath it as a termite barrier.

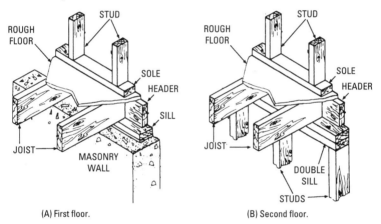

(A) First floor.   (B) Second floor.

Figure 7-3   Details of platform framing of sill plates and joists for western frame box-sill construction.

T-Sills
There are two types of *T-sill* construction, one commonly used in the South (or in a dry, warm climate), and one used in the North and East (where it is colder). Their construction is similar, except in the East, where the T-sills and joists are nailed directly to the studs, as well as to the sills. The headers are nailed in between the floor joists.

Braced Framing Sills

With *braced framing sills*, the floor joists are notched out and nailed directly to the sills and studs.

Built-Up Sills

Where *built-up sills* are used, the joints are staggered. If piers are used in the foundation, heavier sills are used. These sills are of single heavy timber or are built up of two or more pieces of timber. Where heavy timber or built-up sills are used, the joints should occur over the piers. The size of the sill depends on the load to be carried and on the spacing of the piers. Where earth floors are used, the studs are nailed directly to the sill plates.

For two-story buildings (and especially in locations subject to earthquakes or tornadoes), a double sill is desirable, because it affords a larger nailing surface for sheathing brought down over the sill, and ties the wall framing more firmly to its sills. In cases where the building is supported by posts or piers, it is necessary to increase the sill size, since the sill supported by posts acts as a girder.

Since it is not necessary that the sill be of great strength in most types of construction, the foundation will provide uniformly solid bearing throughout its entire length. The main requirements are:

- Resistance to crushing across the grain
- Ability to withstand decay and attacks of insects
- Availability of adequate nailing area for studs, joists, and sheathing

Anchorage of Sill

It is important (especially in locations with strong winds) that buildings be thoroughly anchored to the foundation (see Figure 7-4). Anchoring is accomplished by setting at suitable intervals (6 to 8 feet) $1/2$-inch bolts that extend at least 18 inches into the foundation. They should project above the sill to receive a good-sized washer and nut. With hollow tile, concrete blocks, and material of cellular structure, particular care should be taken in filling the cells in which the bolts are placed solidly with concrete.

Setting the Sills

After the girder is in position, the sills are placed on top of the foundation walls, are fitted together at the joints, and leveled throughout. The last operation can be done either by a sight level or by laying them in a full bed of mortar and leveling them with the anchor bolts (see Figure 7-4). On the other hand, it can be left loose and then all the bolts can be tightened as needed to bring the sill level.

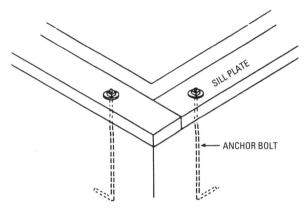

Figure 7-4 One way to anchor sill to foundation. *(Courtesy of the National Forest Products Assn.)*

Sills that are to rest on a wall of masonry should be pressure treated, or kept up at least 18 inches above the ground, as decaying sills are a frightful source of trouble and expense in wooden buildings. Sheathing, paper, and siding should, therefore, be very carefully installed to exclude all wind and wet weather.

Floor Framing
After the girders and sills have been placed, the next operation consists in sawing to size the floor beams or joists of the first floor, and placing them in position on the sills and girders. If there is a great variation in the size of timbers, it is necessary to cut the joists $^1/_2$ inch narrower than the timber so that their upper edges will be in alignment. This sizing should be made from the top edge of the joist (see Figure 7-5). When the joists have been cut to the correct dimension, they should be placed upon the sill and girders, and spaced 16 inches between centers, beginning at one side or end of a room. This is done to avoid wasting material.

Connecting Joist to Sills and Girders
Joists can be connected to sills and girders by several methods, but the prime consideration, of course, is to be sure that the connection is able to hold the load that the joists will carry.

The placing of the girders is an important factor in making the connection. The joists must be level. Therefore, if the girder is not the same height as the sill, the joists must be notched. In placing joists, always have the crown up, since this counteracts the weight

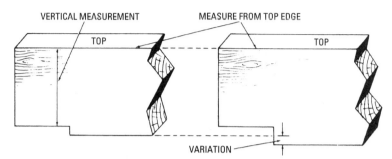

VERTICAL MEASUREMENT MEASURE FROM TOP EDGE

TOP   TOP

VARIATION

Figure 7-5 Showing the variation of joist width.

on them. In most cases, there will be no sag below a straight line. When a joist is to rest on plates or girders, the joist is cut long enough to extend the full width of the plate or girder.

Bridging
To prevent joists from springing sideways under load (which would reduce their carrying capacity), they are tied together diagonally by 1 × 3 or 2 × 3 strips, a process called *bridging* (see Figure 7-6). The 1 × 3 ties are used for small houses, and the 2 × 3 stock on larger work. Metal bridging may also be used (see Figure 7-7).

Rows of bridging should not be more than 8 feet apart. Bridging pieces may be cut all in one operation with a miter box, or the bridging may be cut to fit. Bridging is put in before the subfloor is laid, and each piece is fastened with two nails at the top end. The subfloor should be laid before the bottom end is nailed.

A more-rigid (less-vibrating) floor can be made by cutting in solid 2-inch joists of the same depth. They should be cut perfectly square and a little full (say, $1/16$ inch longer than the inside distance between the joists). First, set a block in every other space, then go back and put in the intervening ones. This keeps the joists from spreading and allows the second ones to be driven in with the strain the same in both directions. This solid blocking is much more effective than cross bridging (see Figure 7-8). The blocks should be toe nailed, and not staggered and nailed through the joists.

Headers and Trimmers
The foregoing operations would complete the first-floor framework in rooms having no framed openings (such as those for stairways, chimneys, and elevators).

The definition of a *header* is a short transverse joist that supports the ends of one or more joists where they are cut off at an

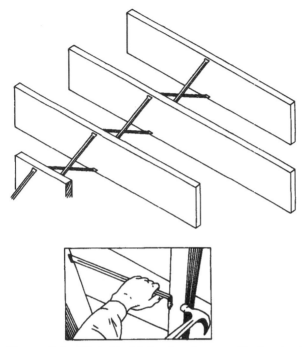

Figure 7-6   Metal bridging in place and being installed.

CROSS BRIDGING

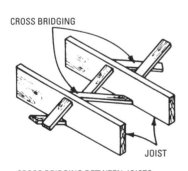

JOIST

CROSS BRIDGING BETWEEN JOISTS

Figure 7-7   Cross bridging between floor joists made of 1-inch × 3-inch wood.

opening. A *trimmer* is a carrying joist that supports an end of a header.

Figure 7-9 shows typical floor openings used for chimneys and stairways. For these openings, the headers and trimmers are set in place first. The floor joists are installed next. Metal hangers are available.

Figure 7-8 Solid bridging between joists.

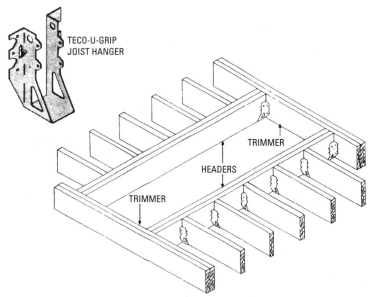

Figure 7-9 One of many framing fasteners available. These are *joist hangers,* **but are commonly referred to as** *Tecos,* **after the brand name.**

Headers run at right angles to the direction of the joists. They are doubled. Trimmers run parallel to the joists and are actually doubled joists. The joists are framed to the headers where the headers form the opening frame at right angles to the joists. These shorter joists framed to the headers are called *tail beams*. The number of headers and trimmers required at any opening depends upon the shape of the opening.

Subflooring

With the sills and floor joists completed, it is necessary to install the subflooring. The subflooring is permanently laid before erecting any wall framework, since the wall plate rests on it (see Figure 7-10). This floor is called the *rough floor* (or *subfloor*, or *deck*) and may be viewed as a large platform covering the entire width and length of the building. Two layers or coverings of flooring material (subflooring and finished flooring) are placed on the joists. The subfloor may be 1-inch × 4-inch square-edge stock, 1 × 6 or 1 × 8 shiplap, or 4-foot × 8-foot sheets of plywood.

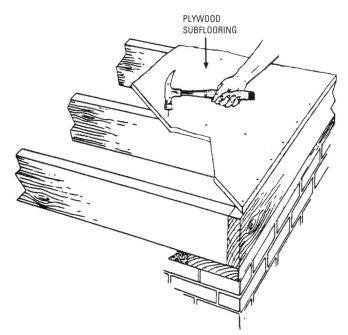

PLYWOOD
SUBFLOORING

Figure 7-10 Board subflooring is laid diagonally for greater strength.

A finished wood floor is in most cases $3/4$-inch tongue-and-groove hardwood (such as oak). Prefinished flooring can also be obtained. In addition, there are nonwood materials, which will be covered in detail later in this book.

Summary

A girder is a principal beam extending from wall to wall of a building to support the joists or floor beams where the distance is too great for a single span. Girders can be made of wood, which must be straight and free of knots. Girders are made three, four, five, or six pieces thick, depending on the load per linear foot and length of the girder.

A sill is the part of the sidewalls of a house that rests horizontally upon, and is securely fastened to, the foundation. The sills are the first part of the framing to be set in place. Therefore, it is important that the sill be constructed properly and placed properly on the foundation. There are various types of sills used; which sill depends on the type of house construction. The size of the sill depends on the load to be carried and on the spacing of the piers. In some construction, two sill plates are used, one nailed on top of the other.

After the sill plate and girder have been constructed and installed, the next operation is to install the floor joists. The joists can be installed in a variety of ways.

To prevent joists from warping under load (which would reduce their carrying capacity), they are tied together diagonally with 1 × 3 or 2 × 3 boards or metal bridging. Bridging is installed before the subflooring is laid. The subfloor should be laid before the bottom end is nailed.

After the joists are installed, the subflooring is laid. In most cases, board subfloor is laid diagonally to give strength and prevent squeaks in the floor. The material generally used for subfloors is 1 × 6 or 1 × 8 boards, or 4-foot × 8-foot plywood panels.

The finished floor in most cases is $1/2$-inch tongue-and-groove hardwood. To save time and labor, pre-finished flooring can be obtained. Where heavy loads are to be carried on the floor, 2-inch flooring should be used.

Review Questions

1. What must be done to align the top edge of floor joists?
2. What is bridging, and how is it installed?
3. How should subflooring be laid?
4. When are headers and trimmers used in floor joists?

5. What are tail beams?

6. Why are girders used?

7. How are sills anchored to the foundation?

8. Name the various types of sills constructed.

9. What is the purpose of the sill?

10. How are engineered girders made?

Chapter 8

Constructing Walls and Partitions

Constructing walls and partitions involves knowing something about bracing and working with studs. Partitions must be plumb and square, while corners must be properly aligned and braced. Erecting the frame and making it upright and properly braced is also important in the construction of any building.

Built-Up Corner Posts

Corner posts may be built up in many ways. You can use studding or larger-sized pieces. Some carpenters form corner posts with two 2-inch × 4-inch studs spiked together to make one piece having a 4 × 4 section. This is suitable for small structures with no interior finish. Figure 8-1 shows various arrangements of built-up posts commonly used.

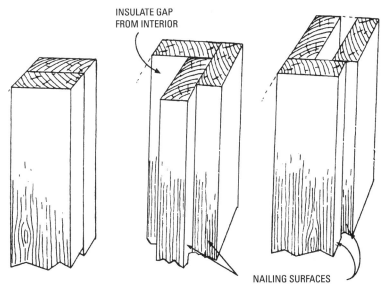

INSULATE GAP
FROM INTERIOR

NAILING SURFACES

Figure 8-1   Various ways to build up corner post.

Bracing

There are two kinds of permanent bracing. One type is called *cut-in bracing* (see Figure 8-2). A house braced in this manner withstood one of the worst hurricanes ever to hit the eastern seaboard, and engineers, after an inspection, gave the bracing the full credit for the survival. The other type of bracing is called *plank bracing* (see Figure 8-3). This type is put on from the outside, the studs being cut and notched so that the bracing is flush with the outside edges of the studs. This method of bracing is commonly used and is very effective. Usually bracing is not needed when plywood is used as sheathing.

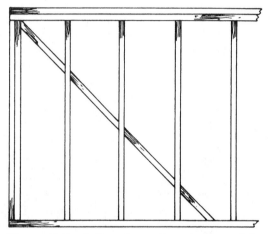

Figure 8-2   A 2 x 4 cut between stud bracing. This type of bracing is called *cut-in*.

Preparing the Corner Posts and Studding

Studs are cut on a radial-arm saw with a stop clamped to the fence to ensure that the studs are all the same length.

Erecting the Frame

If the builder is shorthanded, or working alone, a frame can be crippled together, one member at a time. An experienced carpenter can erect the studs and toenail them in place with no assistance whatever. Then the corners are plumbed and the top plates nailed on from a stepladder. Where enough workers are available, most

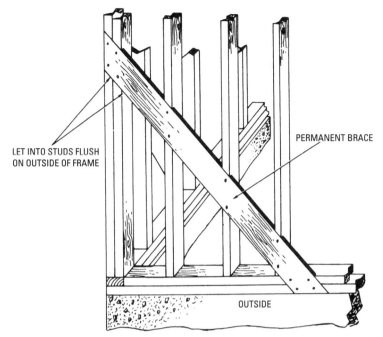

LET INTO STUDS FLUSH
ON OUTSIDE OF FRAME

PERMANENT BRACE

OUTSIDE

Figure 8-3   Plank-type bracing.

contractors prefer to nail the sole plates and top plates while the entire wall is lying on the subfloor. Some even cut all the door and window openings; after that, the entire gang raises the assembly and nails it in position. This is probably the speediest of any possible method, but it cannot be done by only one or two workers. Some contractors even put on the outside sheathing before raising the wall, and then use a lift truck or a high-lift excavating machine to raise it into position.

Framing Around Openings

The openings should be laid out and framed complete. Studding at all openings must be double to furnish more area for proper nailing of the trim, and it must be plumb (see Figure 8-4).

It is necessary that some parts of the studs be cut out around windows or doors in outside walls or partitions. It is imperative to insert some form of a header to support the lower ends of the top studs that have been cut off. A member termed a *rough sill* is located

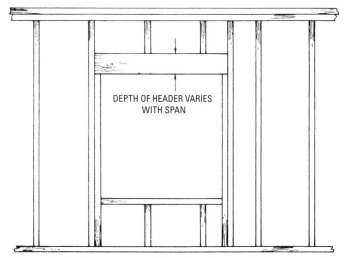

Figure 8-4   An approved framing for a window opening.

at the bottom of the window openings. This sill serves as a nailer, but does not support any weight.

Headers

Headers (see Figure 8-5) are of the following two classes:

- Nonbearing headers occur in walls that are parallel with the joists of the floor above and carry only the weight of the framing immediately above.

- Load-bearing headers occur in walls that carry the end of the floor joists on plates or rib bands immediately above the openings. They must, therefore, support the weight of the floor or floors above.

Size of Headers

The determining factor in header sizes is whether they are load-bearing. In general, it is considered good practice to use a double 2 × 4 header placed on edge unless the opening in a nonbearing partition is more than 3 feet wide. In cases where the trim inside and outside is too wide to prevent satisfactory nailing over the openings, it may become necessary to double the header to provide a nailing base for the trim.

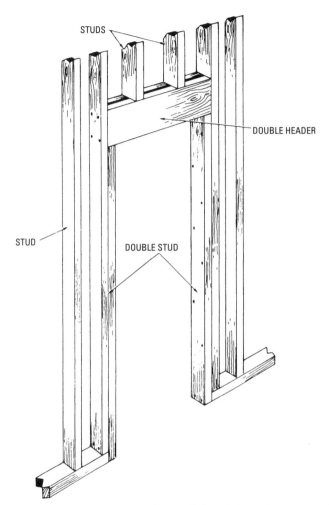

Figure 8-5 Construction of headers when openings over 30 inches between studs appear in partitions or outside walls.

Opening Sizes for Windows and Doors

It is common to frame the openings with dimensions 2 inches more than the unit dimension. This will allow for the plumbing and leveling of the window. For the rough opening of stock windows, check the window book from your supplier. For custom windows, the rough opening should be specified on the drawings. If not, wait until the windows are on-site to get exact measurements.

Interior Partitions

Interior walls that divide the inside space of the buildings into rooms or halls are known as *partitions*. These are made up of studding covered with plasterboard and plaster, metal lath and plaster, or drywall (see Figure 8-6).

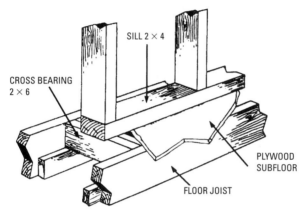

Figure 8-6   A method of constructing a partition between two floor joists.

An *interior partition* differs from an *outside partition* in that it seldom rests on a solid wall. Its support, therefore, requires careful consideration, making sure it is large enough to carry the required weight. The various interior partitions may be bearing or nonbearing. They may run at right angles or parallel to the joists upon which they rest.

Partitions Parallel to Joists

Here the entire weight of the partition will be concentrated on one or two joists, which perhaps are already carrying their full shares of the floor load. In most cases, additional strength should be provided. One method is to provide double joists under such partitions (that is, to put an extra joist beside the regular ones). Computation shows that the average partition weighs nearly three times as much as a single joist should be expected to carry. The usual (and approved) method is to double the joists under nonbearing partitions. An alternative method is to place a joist on each side of the partition.

Where partitions are placed between and parallel to floor joists, bridges must be placed between the joists to provide a means of fastening the partition plate.

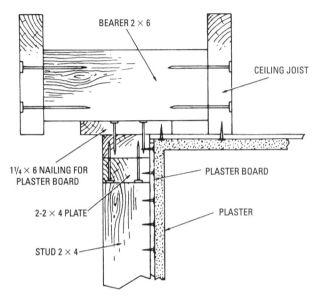

BEARER 2 × 6

CEILING JOIST

1¼ × 6 NAILING FOR
PLASTER BOARD

PLASTER BOARD

2-2 × 4 PLATE

PLASTER

STUD 2 × 4

Figure 8-7   A method of constructing a partition between two ceiling joists.

Figure 8-7 shows the construction at the top. Openings over 30 inches wide in partitions or outside walls must have heavy headers, as shown in Figure 8-5. The partition wall studs are arranged in a row with their ends bearing on a long horizontal member called a *bottom plate*, and their tops capped with another plate called a *top plate*. Double top plates are used in bearing walls and partitions. The bearing strength of stud walls is determined by the strength of the studs.

Partitions at Right Angles to Joists
For nonbearing partitions, it is not necessary to increase the size or number of the joists. The partitions themselves may be braced, but even without bracing, they have some degree of rigidity.

Engineered Wood and I-Joist Open Metal Web System
The patented SpaceJoist TE combines the best features of an I-joist and the SpaceJoist open metal-web system. It creates a different

type of floor joist. It can be cut off at the job site, which gives the installer flexibility. This, combined with the excellent shear and bearing performance of I-joists, makes the SpaceJoist TE the near-perfect floor joist concept.

The open-web system allows easy passage of ductwork, plumbing, and electrical wiring within the floor. No cutting of plywood webs is required, and no furring down is necessary to hide mechanicals (see Figure 8-8).

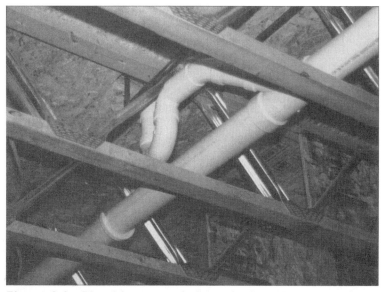

Figure 8-8   Plumbing routed through the metal web of the joist. *(Courtesy of Truswal Systems Corp.)*

The webs are made of high-strength galvanized steel. The Space-Joist TE is built with reliable stress-graded lumber. Each web has integrally formed metal teeth that are pressed into the sides of the chords. Each metal web has a patented deep-shape reinforcing rib that allows the web to be used in compression or tension.

Labor and Material Costs Reduction
On-site, in-place costs are competitive with conventional joist or truss systems. The SpaceJoist TE design provides a lightweight, consistent, quality joist for fast placement at any job, dramatically reducing on-site labor costs. On-flat construction gives wide ($2^1/_2$-inch) surfaces in order to speed gluing and nailing the floor

sheathing. The joist is custom-engineered using sophisticated computer design software, creating a product that is structurally superior to conventional framing, far outperforming dimensional lumber. Designs are checked and sealed by registered Professional Engineers (PEs).

The open-web design and variety of depths allow placement of 12 inches or more of insulation without condensation problems, and eliminates the need for special air-circulation devices. The designer or architect has freedom to create unique, modern structures, unhampered by the span limitations of conventional floor joists. The shop-fabricated joists offer a range of depths that provide long, clear spans. The noncombustible webs eliminate a large portion of the combustible material usually found in the floor. A variety of fire-endurance assemblies are available to meet fire-rating requirements.

The newer-type engineered-wood joists serve a number of purposes and improve home safety. It also increases weight-bearing loads that were not available in conventional joists, as well as offering easy installation. The following illustrations will bear out the advantages of the latest in construction techniques that add to the value and soundness of new structures.

Figure 8-9 shows how the portable power handsaw is used to cut the ends of the joists to fit the span. Figure 8-10 presents a variety of applications, with depths ranging from $9^{1}/_{4}$ inches to $15^{3}/_{4}$ inches.

Figure 8-9   Using a power handsaw to cut joist ends to fit. *(Courtesy of Truswal Systems Corp.)*

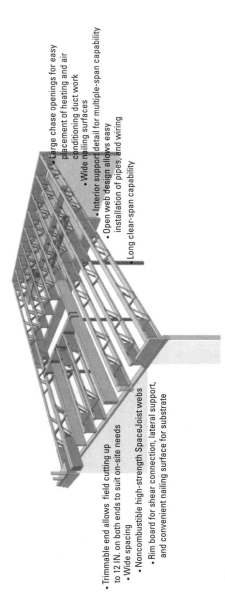

Figure 8-10 Joists mounted over a basement. *(Courtesy of Truswal Systems Corp.)*

- Large chase openings for easy placement of heating and air conditioning duct work
- Wide nailing surfaces
- Interior support detail for multiple-span capability
- Open web design allows easy installation of pipes, and wiring

Long clear-span capability

- Trimmable end allows field cutting up to 12 IN. on both ends to suit on-site needs
- Wide spacing
- Noncombustible high-strength SpaceJoist webs
- Rim board for shear connection, lateral support, and convenient nailing surface for substrate

Figure 8-11 Crawl-space construction on a slab with Space-Joist TE joists in place. *(Courtesy of Truswal Systems Corp.)*

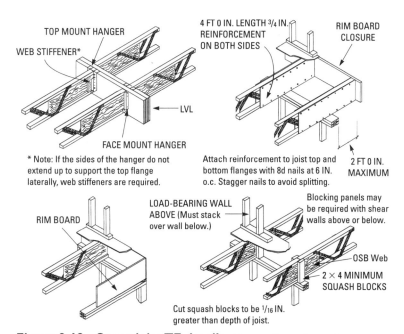

TOP MOUNT HANGER

WEB STIFFENER*

LVL

FACE MOUNT HANGER

4 FT 0 IN. LENGTH ³/₄ IN.
REINFORCEMENT
ON BOTH SIDES

RIM BOARD
CLOSURE

* Note: If the sides of the hanger do not
extend up to support the top flange
laterally, web stiffeners are required.

Attach reinforcement to joist top and
bottom flanges with 8d nails at 6 IN.
o.c. Stagger nails to avoid splitting.

2 FT 0 IN.
MAXIMUM

RIM BOARD

LOAD-BEARING WALL
ABOVE (Must stack
over wall below.)

Blocking panels may
be required with shear
walls above or below.

OSB Web

2 × 4 MINIMUM
SQUASH BLOCKS

Cut squash blocks to be ¹/₁₆ IN.
greater than depth of joist.

Figure 8-12   Space joist TE details. *(Courtesy of Truswal Systems Corp.)*

Table 8-1   Maximum Dimensions for Joists in Inches

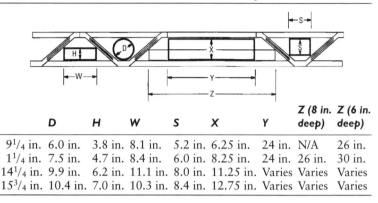

D	H	W	S	X	Y	Z (8 in. deep)	Z (6 in. deep)
9¹/₄ in.	6.0 in.	3.8 in.	8.1 in.	5.2 in.	6.25 in.	24 in.	N/A	26 in.
1¹/₄ in.	7.5 in.	4.7 in.	8.4 in.	6.0 in.	8.25 in.	24 in.	26 in.	30 in.
14¹/₄ in.	9.9 in.	6.2 in.	11.1 in.	8.0 in.	11.25 in.	Varies	Varies	Varies
15³/₄ in.	10.4 in.	7.0 in.	10.3 in.	8.4 in.	12.75 in.	Varies	Varies	Varies

Table 8-2   SpaceJoist TE Floor Span Chart

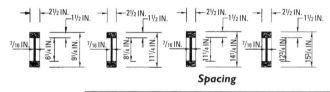

Depth	Deflection	Spacing 24 in. o.c.	19.2 in. o.c.	16 in. o.c.	12 in. o.c.
9¹/₄ in.	L/480	14 ft 3 in.	14 ft 11.5 in.	15 ft 9 in.	16 ft 10.5 in.
	L/360	15 ft 9 in.	16 ft 7 in.	17 ft 4 in.	19 ft 0 in.
11¹/₄ in.	L/480	16 ft 4.5 in.	17 ft 3 in.	18 ft 4 in.	19 ft 11 in.
	L/360	17 ft 10.5 in.	19 ft 4 in.	20 ft 0 in.	20 ft 0 in.
14¹/₄ in.	L/480	18 ft 0 in.	20 ft 0 in.	20 ft 10 in.	22 ft 0 in.
	L/360	18 ft 0 in.	21 ft 6 in.	22 ft 0 in.	22 ft 0 in.
15³/₄ in.	L/480	20 ft 0 in.	21 ft 6 in.	22 ft 10 in.	24 ft 0 in.
	L/360	20 ft 0 in.	22 ft 0 in.	24 ft 0 in.	26 ft 0 in.

40# PSF Live Load. 10# PSF T.C. Dead Load. 5# PSF B.C. Dead Load + 55# PSF Total Load

Notes:

Up to 12 inches on both ends are trimmable.

Spans shown are based on a floor loading of 40 psf live load and 15 psf dead load (10 psf T.C., 5 psf B.C.)

Spans shown are out-to-out dimensions and include the bearing length. 1³/₄-inch minimum bearing is required at joist ends.

Spans shown assume composite action with single layer of the appropriate span-rated, glue-nailed wood sheathing for deflection only.

In addition, Figure 8-11 illustrates how the joists are placed in a crawl-space type of construction.

Table 8-1 shows maximum dimensions for the joists. Table 8-2 shows the floor span and spacing other than 24-inch and 16-inch on-center. Note the 19.2-inch and 12-inch spacing. Figure 8-12 shows details of the joists and how they are put to work in a building.

Summary

Bracing is a very important part of outer-wall framing, and there are generally two types: cut-in and plank.

A carpenter who is working alone can erect each stud and toenail into the sole plate with no assistance whatsoever. The top plate can be nailed from a stepladder, after the corners and studs are plumb. When enough workers are available, most contractors prefer to nail the sole plates and top plates as the entire wall is lying on the floor or ground. In many cases, all of the door and window openings are constructed and nailed in place. Then the wall is raised as one unit and nailed in position.

Interior walls that divide the inside space into rooms or halls are generally known as partitions. They are made up of studding with each joist set 16 inches on-center. Where partitions are to be placed between and parallel to the floor joists, bridges must be placed between the joists to provide a means of fastening the partition plate.

Review Questions

1. Why is corner bracing so important?
2. What is the difference between a load-bearing and a nonload-bearing partition?
3. What is a bearing member?
4. Why should double headers and studs be installed in door and window openings?
5. Why are corner-post designs so important?
6. Name the two kinds of permanent bracing.
7. How are studs cut?
8. What is a header?
9. How are headers classified?
10. How do you reinforce overloaded floor joists?

Chapter 9

Framing Roofs

As a preliminary to the study of this chapter, you should review Chapter 17, "Using the Steel Square," in the book, *Audel Carpenters and Builders Tools, Steel Square, and Joinery* (Wiley Publishing, Inc., 2004) in this series of books (see the Introduction for details on the series). This tool, which is invaluable to the carpenter in roof framing, has been explained in detail in that chapter, with many examples of rafter cutting. Hence, knowledge of how to use the square is assumed here to avoid repetition.

Types of Roofs

Following are some of the many types of roofs used in construction:

- *Shed or lean-to roof*—This is the simplest and least-expensive form of roof and is usually employed for small sheds and out-buildings (see Figure 9-1). It has a single slope.

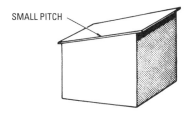
SMALL PITCH

Figure 9-1 Shed or lean-to roof used on small sheds or buildings.

- *Gable or pitch roof*—This is a very common, simple, and efficient form of roof and is used extensively on all kinds of buildings. It is of triangular section, having two slopes meeting at the center or ridge and forming a gable (see Figure 9-2). It is popular because of the ease of construction, relative economy, and efficiency.

- *Gambrel roof*—This is a modification of the gable roof, each side having two slopes (see Figure 9-3).

- *Hip roof*—A hip roof is formed by four straight sides, all sides sloping toward the center of the building, and terminating in a ridge instead of a deck (see Figure 9-4).

- *Hip-and-valley roof*—This is a combination of a hip roof and an intersecting gable, so-called because both hip and valley rafters are required in its construction. There are many

Figure 9-2   Gable or pitch roof. *(Courtesy of Shetter-Kit, Inc.)*

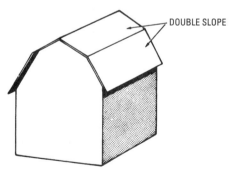

Figure 9-3   Gambrel roof.

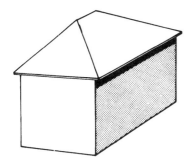

Figure 9-4   Hip roof.

modifications of this roof. Usually the intersection is at right angles, but it need not be. Either ridge may rise above the other, and the pitches may be equal or different, thus giving rise to an endless variety (see Figure 9-5).

- *Mansard roof*—The straight sides of this roof slope vary steeply from each side of the building toward the center, and the roof has a nearly-flat deck on top (see Figure 9-6).
- *French or concave mansard roof*—This is a modification of the Mansard roof, its sides being concave instead of straight (see Figure 9-7).

Roof Construction

The frames of most roofs are made up of timbers called *rafters*. These are inclined upward in pairs, their lower ends resting on the top plate, and their upper ends being tied together with a ridge board. On large buildings, such framework is usually reinforced by interior supports to avoid using abnormally large timbers.

The prime objective of a roof in any climate is to keep out water. The roof must be sloped or otherwise built to shed water. Where heavy snows cover the roof for long periods, it must be constructed more rigidly to bear the extra weight. Roofs must also be strong enough to withstand high winds.

Following are terms used in connection with roofs:

- *Span*—The *span* of any roof is the shortest distance between the two opposite rafter seats. Stated another way, it is the measurement between the outside plates, measured at right angles to the direction of the ridge of the building.
- *Total rise*—The *total rise* is the vertical distance from the plate to the top of the ridge.

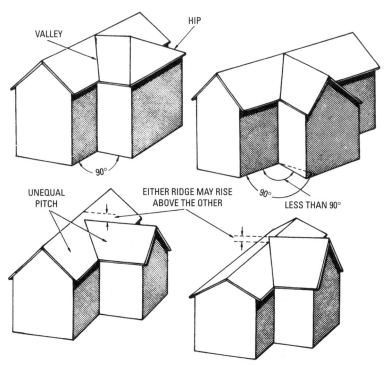

Figure 9-5   Various styles of hip-and-valley roofs.

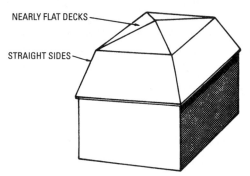

Figure 9-6   Mansard roof.

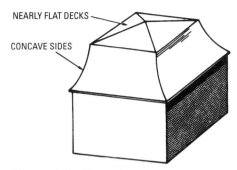

NEARLY FLAT DECKS

CONCAVE SIDES

Figure 9-7   French or concave Mansard roof.

- *Total run*—The term *total run* always refers to the level distance over which any rafter passes. For the ordinary rafter, this would be one-half the span distance.
- *Unit of run*—The unit of measurement (1 foot or 12 inches) is the same for the roof as for any other part of the building. By the use of this common unit of measurement, the framing square is employed in laying out large roofs.
- *Rise in inches*—The *rise in inches* is the number of inches that a roof rises for every foot of run.
- *Pitch*—*Pitch* is the term used to describe the amount of slope of a roof.
- *Cut of roof*—The *cut of a roof* is the rise in inches and the unit of run (12 inches).
- *Line length*—The term *line length* as applied to roof framing is the hypotenuse of a triangle whose base is the total run and whose altitude is the total rise.
- *Plumb and level lines*—These terms have reference to the direction of a line on a rafter, and not to any particular rafter cut. Any line that is vertical when the rafter is in its proper position is called a *plumb line*. Any line that is level when the rafter is in its proper position is called a *level line*.

Rafters

Rafters are the supports for the roof covering and serve in the same capacity as joists do for the floor or studs do for the walls. Rafters are sized according to the distance they must span and the load they must carry.

The carpenter should thoroughly know these various types of rafters, and be able to distinguish each kind as they are briefly described. Following are the various kinds of rafters used in roof construction:

- *Common rafter*—An example of a common rafter is one extending at right angles from plate to ridge (see Figure 9-8).

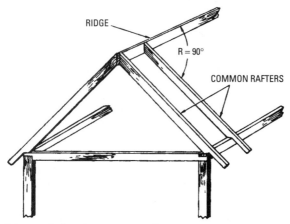

Figure 9-8 Common rafters.

- *Hip Rafter*—An example of a hip rafter is one extending diagonally from a corner of the plate to ridge (see Figure 9-9).
- *Valley rafter*—A rafter that extends diagonally from the plate to the ridge at the intersection of a gable extension and the main roof.
- *Jack rafter*—Any rafter that does not extend from the plate to the ridge is a jack rafter.
- *Hip-jack rafter*—A hip-jack rafter extends from the plate to a hip rafter at an angle of 90° to the plate (see Figure 9-9).
- *Valley-jack rafter*—A valley jack rafter extends from a valley rafter to the ridge at an angle of 90° to the ridge (see Figure 9-10).
- *Cripple-jack rafter*—A cripple-jack rafter extends from a valley rafter to hip rafter and at an angle of 90° to the ridge (see Figure 9-11).

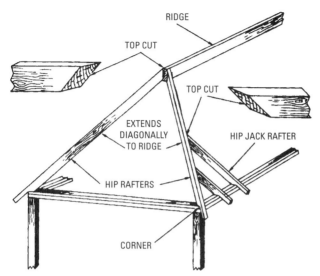

Figure 9-9   Hip-roof rafters.

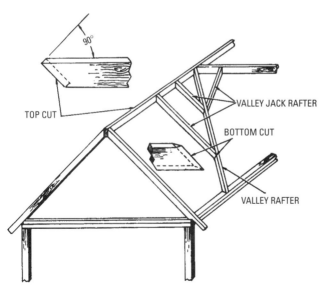

Figure 9-10   Valley and valley-jack rafters.

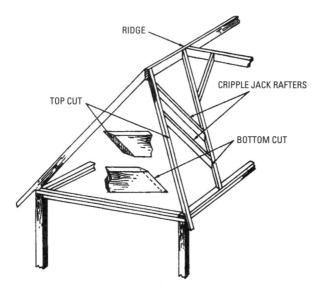

RIDGE

CRIPPLE JACK RAFTERS

TOP CUT

BOTTOM CUT

Figure 9-11   Cripple jack rafters.

- *Octagon rafter*—An octagon rafter is any rafter that extends from an octagon-shaped plate to a central apex, or ridgepole.

A rafter usually consists of a main part or rafter proper, and a short length called the *tail*, which extends beyond the plate. The rafter and its tail may be all in one piece, or the tail may be a separate piece nailed onto the rafter.

Length of Rafter

The length of a rafter may be found in several ways:

- By calculation
- By steel framing square (including the multiposition method, by scaling, and by aid of the framing table)
- By full-scale layout on the deck

Example

What is the length of a common rafter having a run of 6 feet and rise of 4 inches per foot?

By calculation (see Figure 9-12):

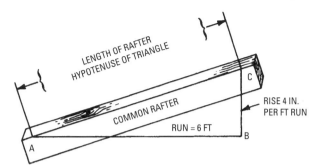

Figure 9-12   Method of finding the length of a rafter by calculation.

The total rise $= 6 \times 4 = 24$ inches $= 2$ feet.

Since the edge of the rafter forms the hypotenuse of a right triangle (whose other two sides are the run and rise) then the length of the rafter (see Figure 9-12) is calculated as follows:

$$= \sqrt{run^2} + \sqrt{rise^2} = \sqrt{6^2 + 2^2} = \sqrt{40} = 6.33 \text{ feet.}$$

Practical carpenters would not consider it economical to find rafter lengths in this way because it takes too much time and there is a chance of error. It is to avoid both objections that the framing square has been developed.

With steel framing square:

The steel framing square considerably reduces the mental effort and chances of error in finding rafter lengths. An approved method of finding rafter lengths with the square is with the aid of the rafter table included on the square for that purpose. However, some carpenters may possess a square, which does not have rafter tables. In such case, the rafter length can be found either by the *multiposition* method (see Figure 9-13), or by *scaling* (see Figure 9-14). In either of these methods, the measurements should be made with care because, in the multiposition method, a separate measurement must be made for each foot run with a chance for error in each measurement.

Problem 1

Lay off the length of a common rafter having a run of 6 feet and a rise of 4 inches per foot. Locate a point A on the edge of the rafter, leaving enough stock for a lookout, if any is used. Place the steel

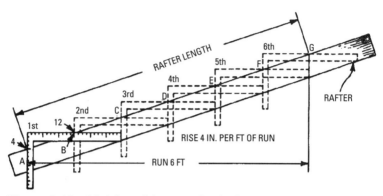

Figure 9-13   Multi-position method of finding rafter length.

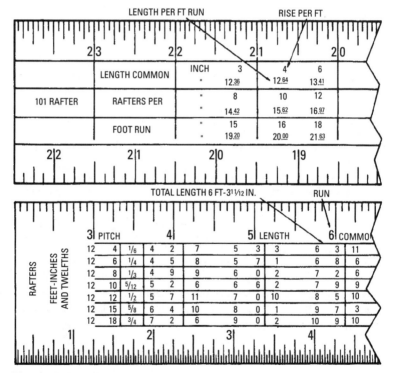

Figure 9-14   Rafter-table readings of two well-known makers of steel framing squares.

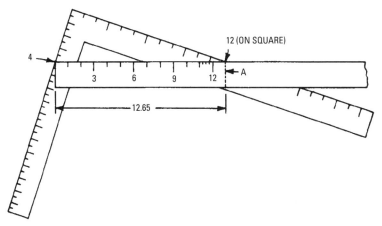

Figure 9-15   Method of finding rafter length by scaling.

framing square so that division 4 coincides with *A*, and 12 registers with the edge of *B*. Evidently, if the run were 1 foot, distance *AB* thus obtained would be the length of the rafter *per foot run*. Apply the square six times, for the 6-foot run, obtaining points *C*, *D*, *E*, *F*, and *G*. The distance *AG*, then, is the length of the rafter for a given run.

Figure 9-14 shows readings of rafter tables of two well-known makes of squares for the length of the rafter in the preceding example, one giving the length per foot run, and the other the total length for the given run.

Problem 2

Given the rise per foot in inches, use two squares, or a square and a straightedge scale (see Figure 9-15). Place the straightedge on the square to be able to read the length of the diagonal between the rise of 4 inches on the tongue and the 1-foot (12-inches) run on the body as shown. The reading is a little more than 12 inches. To find the fraction, place dividers on 12 and at point *A* (see Figure 9-16). Transfer to the hundredths scale and read .65, (see Figure 9-17), making the length of the rafter 12.65 inches *per foot run*, which for a 6-foot run is as follows:

$$\frac{12.65 \times 6}{12} = 6.33 \text{ feet.}$$

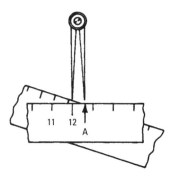

Figure 9-16 Reading the straightedge in combination with the carpenter's square.

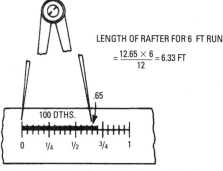

LENGTH OF RAFTER FOR 6 FT RUN

$$= \frac{12.65 \times 6}{12} = 6.33 \, \text{FT}$$

Figure 9-17 Method of reading hundredths scale.

Problem 3

To find the total rise and run given in feet, let each inch on the tongue and body of the square equal 1 foot. The straightedge should be divided into inches and twelfths of an inch so that on a scale, 1 inch = 1 foot. Each division will, therefore, equal 1 inch. Read the diagonal length between the numbers representing the run and rise (12 and 4), taking the whole number of inches as feet, and the fractions as inches. Transfer the fraction with dividers and apply the hundredths scale, as was done in Problem 2 (see Figure 9-16 and Figure 9-17).

In estimating the total length of stock for a rafter having a tail, the run of the tail or length of the lookout must, of course, be considered.

Rafter Cuts

All rafters must be cut to the proper angle or bevel at the points where they are fastened and, in the case of overhanging rafters, also at the outer end. The various cuts are known as:

- Top or plumb
- Bottom, seat, or heel
- Tail or lookout
- Side or cheek

Common Rafter Cuts

All of the cuts for the various types of common rafters are made at right angles to the sides of the rafter (that is, not beveled, as in the case of jacks). Figure 9-18 shows various common rafters from which the natures of these various cuts are seen.

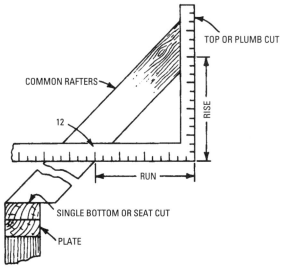

Figure 9-18 Various common rafters illustrating types and names of cuts; showing why one side of the square is set at 12 in laying out the cut.

In laying out cuts from common rafters, one side of the square is always placed on the edge of the stock at 12 (see Figure 9-18). This distance 12 corresponds to 1 foot of the run. The other side of the square is set with the edge of the stock to the rise in inches *per foot run*. This is virtually a repetition of Figure 9-13, but it is very

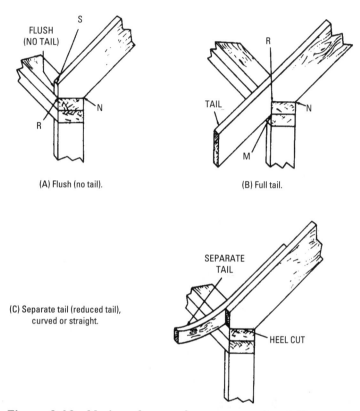

(A) Flush (no tail).

(B) Full tail.

(C) Separate tail (reduced tail), curved or straight.

Figure 9-19   Various forms of common rafter tails.

important to understand why one side of the square is set to 12 for common rafters (that is, not simply to know that 12 must be used). On rafters having a full tail (see Figure 9-19B), some carpenters do not cut the rafter tails, but wait until the rafters are set in place so that they may be lined and cut while in position. Certain kinds of work permit the ends to be cut at the same time the remainder of the rafter is framed.

The method of handling the square in laying out the *bottom* and *lookout cuts* is shown in Figure 9-20. In laying out the *top* or *plumb cut*, if there is a ridge board, one-half of the thickness of the ridge must be deducted from the rafter length. If a lookout or a *tail cut* is to be vertical, place the square at the end of the stock with the rise and run setting (as shown in Figure 9-20), and scribe the cut line *LF*. Lay off *FS* equal to the length of the lookout, and move the square up to *S* (with the same setting) and scribe line *MS*. On this line, lay

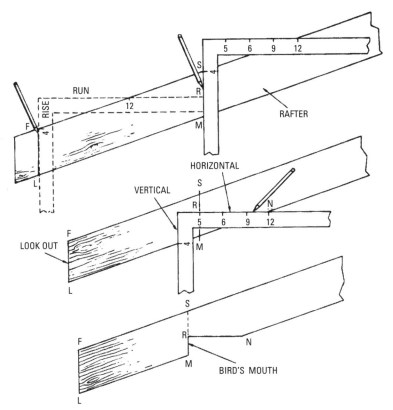

Figure 9-20 Method of using the square in laying out the lower or end cut of the rafter.

off *MR*, the length of the vertical side of the bottom cut. Now apply the same setting to the bottom edge of the rafter, so that the edge of the square cuts *R*, and scribe *RN*, which is the horizontal sideline of the bottom cut. In making the bottom cut, the wood is cut out to the lines *MR* and *RN*. The lookout and bottom cuts are shown made in Figure 9-19B, *RN* being the side that rests on the plate, and *RM* the side that touches the outer side of the plate.

Hip and Valley Rafter Cuts

The hip rafter lies in the plane of the common rafters and forms the hypotenuse of a triangle, of which one leg is the adjacent common rafter, and the other leg is the portion of the plate intercepted between the feet of the hip and common rafters (see Figure 9-21).

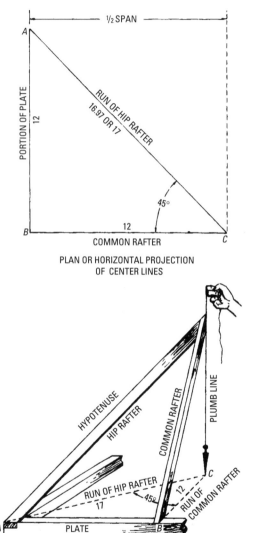

Figure 9-21   View of hip and common rafters in respect to each other.

Problem

In Figure 9-21, take the run of the common rafter as 12, which may be considered as 1 foot (12 inches) of the run, or the total run of 12 feet (half the span). Now, for 12 feet, intercept on the plate the hip run inclined to 45° to the common run, as in the triangle ABC. Thus:

$$AC^2 = \sqrt{AB^2 + BC^2} = \sqrt{12^2 + 12^2}$$

$$= 16.97, \text{ or approximately } 17$$

Therefore, the run of the hip rafter is to the run of the common rafter as 17 is to 12. Accordingly, in laying out the cuts, use figure 17 on one side of the square and the given rise in *inches per foot* on the other side. This also holds true for top and bottom cuts of the valley rafter when the plate intercept AB = the run BC of the common rafter.

The line of measurement for the length of a hip and valley rafter is along the middle of the back or top edge, as on common and jack rafters. The rise is the same as that of a common rafter, and the run of a hip rafter is the horizontal distance from the plumb line of its rise to the outside of the plate at the foot of the hip rafter (see Figure 9-22).

In applying the source for cuts of hip or valley rafters, use the distance 17 on the body of the square in the same way as 12 was used for common rafters. When the plate distance between hip and common rafters is equal to half the span or run of the common rafter, the line of run of the hip will lie at 45° to the line of the common rafter (see Figure 9-21).

The length of a hip rafter (as given in the framing table on the square) is the distance from the ridge board to the outer edge of the plate. In practice, deduct from this length one-half the thickness of the ridge board, and add for any projection beyond the plate for the eave. Figure 9-23A shows the correction for the table length of a hip rafter to allow for a ridge board, and Figure 9-23B shows the correction at the plate end that may or may not be made as in Figure 9-24.

The table length, as read from the square, must be reduced an amount equal to MS. This is equal to the hypotenuse (ab) of the little triangle abc, which in value equals

$$\sqrt{ac^2 + bc^2} = \sqrt{ac^2 \times (\text{half thickness of ridge})^2}.$$

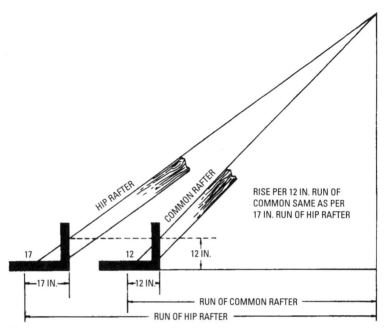

Figure 9-22 Hip and common rafters shown in the same plane. This illustrates the use of 12 for the common rafter and 17 for the hip rafter.

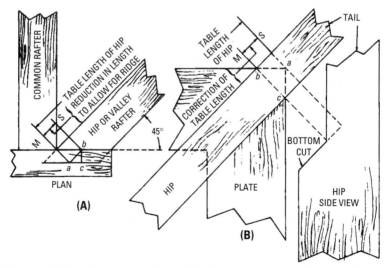

Figure 9-23 Correction in table for top cut to allow for half thickness of ridge board.

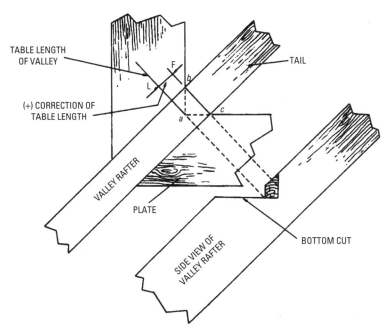

TABLE LENGTH
OF VALLEY

TAIL

(+) CORRECTION OF
TABLE LENGTH

VALLEY RAFTER

PLATE

SIDE VIEW OF
VALLEY RAFTER

BOTTOM CUT

Figure 9-24 Side view of valley rafter showing bottom and seat cut at top plate.

In ordinary practice, take *MS* as equal to half the thickness of the ridge. The plan and side view of the hip rafter shows the table length and the correction *MS*, which must be deducted from the table length so that the sides of the rafter at the end of the bottom cut will intersect the outside edges of the plate. The table length of the hip rafter (as read on the framing square) will cover the span from the ridge to the outside cover *a* of the plate, but the side edges of the hip intersect the plates at *b* and *c*. The distance that *a* projects beyond a line connecting *bc* or *MS* must be deducted (that is, measured backward toward the ridge end of the hip). In making the bottom cut of a valley rafter, it should be noted that a valley rafter differs from a hip rafter in that the correction distance for the table length must be added instead of subtracted, as for a hip rafter. A distance *MS* was subtracted from the table length of the hip rafter (see Figure 9-23B), and an equal distance (*LF*) was added for the valley rafter (see Figure 9-24).

After the plumb cut is made, the end must be mitered outward for a hip (see Figure 9-25) and inward for a valley (see Figure 9-26)

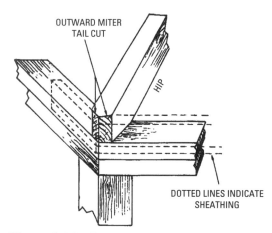

Figure 9-25 Flush hip-rafter miter cut.

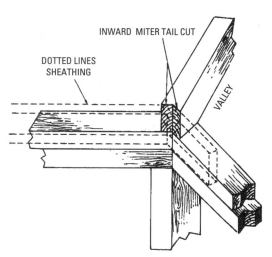

Figure 9-26 Flush valley-rafter miter cut.

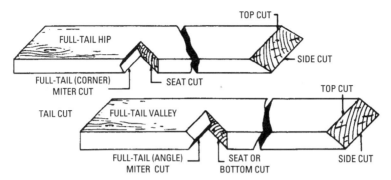

Figure 9-27   Full-tail hip and valley rafters showing all cuts.

to receive the fascia. A *fascia* is the narrow vertical member fastened to the outside ends of the rafter tails. The miter cuts are shown with full tails in Figure 9-27, which illustrates hip and valley rafters in place on the plate.

Side Cuts of Hip and Valley Rafters

These rafters have a side or cheek cut at the ridge end. In the absence of a framing square, a simple method of laying out the side cut for a 45° hip or valley rafter is as follows.

Measure back on the edge of the rafter from point A of the top cut (see Figure 9-28). Distance AC is equal to the thickness of the rafter. Square across from C to B on the opposite edge, and scribe line AB, which gives the side cut. FA is the top cut, and AB is the side cut. Here A, the point from which half the thickness of the rafter is measured, is seen at the top end of the cut. This rule does not hold for any angle other than 45°.

Backing of Hip Rafters

By definition, the term *backing* is the bevel on the top side of a hip rafter that allows the roofing boards to fit the top of the rafter without leaving a triangular hole between it and the back of the roof covering. The height of the hip rafter (measured on the outside surface vertically upward from the outside corner of the plate) will be the same as that of the common rafter measured from the same line, whether the hip is backed or not. This is not true for an unbacked valley rafter when the measurement is made at the center of the timber.

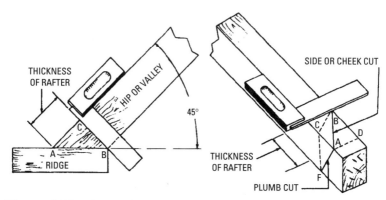

Figure 9-28 A method of obtaining a side cut of 45° hip or valley rafter without aid of a framing table.

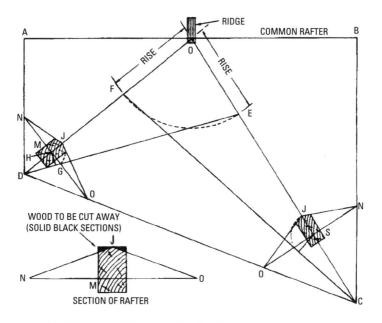

Figure 9-29 Graphical method of finding length of rafters and backing of hip rafters.

Figure 9-29 shows the graphical method of finding the backing of hip rafters. Let *AB* be the span of the building, and *OD* and *OC* the span of two unequal hips. Lay off the given rise as shown. Then *DE* and *CF* are the lengths of the two unequal hips. Take any point, such as *G* on *DE*, and erect a perpendicular cutting *DF* at *H*. Revolve *GH* to *J* (that is, make *HJ = GH)*, draw *NO* perpendicular to *OD* and through *H*. Join *J* to *N* and *O*, giving a bevel angle *NJO*, which is the backing for rafter *DE*. Similarly, the bevel angle *NJO* is found for the backing of rafter *CF*.

Jack Rafters
There are several kinds of jack rafters, and they are distinguished by their relation with other rafters of the roof. These various jack rafters are known (see Figure 9-30) as follows:

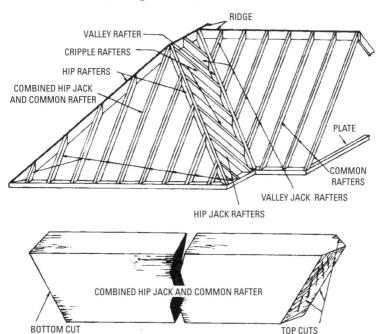

Figure 9-30   A perspective view of hip and valley roof showing the various kinds of jack rafters, and enlarged detail of combined hip-jack and common rafters showing cuts.

• Hip jacks—Rafters that are framed between a hip rafter and the plate.

- Valley jacks—Rafters that are framed between the ridge and a valley rafter.
- Cripple jacks—Rafters that are framed between hip and valley rafters.

The term *cripple* is applied because the ends or *feet* of the rafters are cut off (the rafter does not extend the full length from ridge to plate). From this point of view, a valley jack is sometimes erroneously called cripple. It is virtually a semi-cripple rafter, but confusion is avoided by using the term cripple for rafters framed between the hip and valley rafters.

Jack rafters are virtually discontinuous common rafters. They are cut off by the intersection of a hip or valley (or both) before reaching the full length from plate to ridge. Their lengths are found in the same way as for common rafters (the number 12 being used on one side of the square and the rise in inches per foot run on the other side). This gives the length of jack rafter per foot run, and is true for all jacks (hip, valley, and cripple).

In actual practice, carpenters usually measure the length of hip or valley jacks from the long point to the ridge, instead of along the center of the top, with no reduction being made for one-half the diagonal thickness of the hip or valley rafter. Cripples are measured from long point to long point, with no reduction being made for the thickness of the hip or valley rafter.

Because no two jacks are of the same length, various methods of procedure are employed in framing, including the following:

- Beginning with shortest jack
- Beginning with longest jack
- Using framing table

Shortest-Jack Method
Begin by finding the length of the shortest jack. Take its spacing from the corner, measured on the plates, which, in the case of a 45° hip, is equal to the jack's run. The length of this first jack will be the common difference that must be added to each jack to get the length of the next longer jack.

Longest-Jack Method
Where the longest jack is a full-length rafter (that is, a common rafter), first find the length of the longest jack, then count the spaces between jacks and divide the length of the longest jack by the number of spaces. The quotient will be the common difference. Then frame

the longest jack and make each jack shorter than the preceding jack by this common difference.

Framing-Table Method

On various steel squares, there are tables giving the length of the shortest jack rafters corresponding to the various spacings (such as 16, 20, and 24 inches) between centers for the different pitches. This length is also the common difference and thus serves for obtaining the length of all the jacks.

Example

Find the length of the shortest jack or the common difference in the length of the jack rafters, where the rise of the roof is 10 inches per foot and the jack rafters are spaced 16 inches between centers; also, when spaced 20 inches between centers. Figure 9-31 shows the reading of the jack table on the square for 16-inch centers, and Figure 9-32 shows the reading on the square for 20-inch centers.

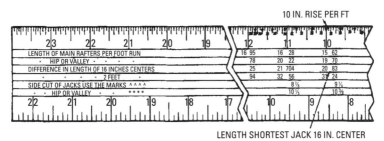

Figure 9-31   Square showing table for shortest jack rafter at 16 inches on center.

Jack-Rafter Cuts

Jack rafters have top and bottom cuts that are laid out the same as for common rafters, and side cuts that are laid out the same as for a hip rafter. To lay off the top or plumb cut with a square, take 12 on the tongue and the rise in inches (of common rafter) per foot run on the blade, and mark along the blade (see Figure 9-33). The following example illustrates the use of the framing square in finding the side cut.

Example

Find the side cut of a jack rafter framed to a 45° hip or valley for a rise of 8 inches per foot run. Figure 9-34 shows the reading on the jack side-cut table of the framing square, and Figure 9-35 shows the method of placing the square on the timber to obtain the side

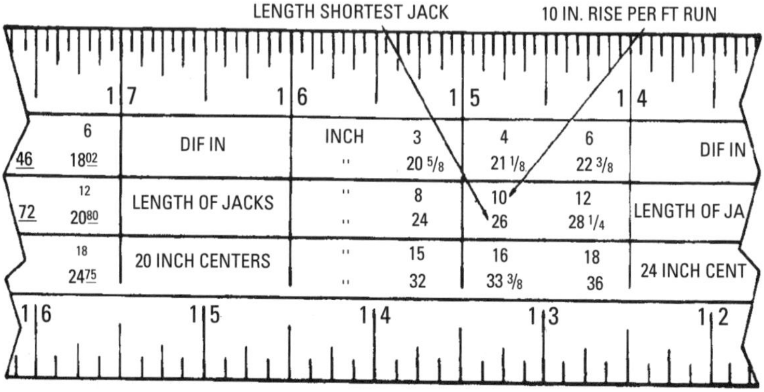

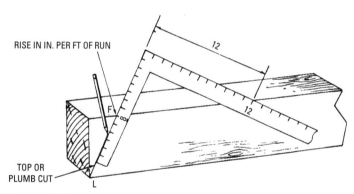

Figure 9-32 Square showing table for shortest jack rafter at 20 inches on center.

Figure 9-33 Method of finding plumb and side cuts of jack framed to 45° hip or valley.

cut. It should be noted that different makers of squares use different setting numbers, but the ratios are always the same.

Method of Tangents

The tangent value is made use of in determining the side cuts of jack, hip, or valley rafters. By taking a circle with a radius of 12 inches, the value of the tangent can be obtained in terms of the constant of the common rafter run.

Considering rafters with zero pitch (see Figure 9-36), if the common rafter is 12 feet long, the tangent *MS* of a 45° hip is the same length. Placing the square on the hip, setting to 12 on the tongue

JACK SIDE CUT · 8 IN. RISE PER FT RUN

1|1				1|0			9			8	
4 25 ¼	6 26 ⅞	FIGURES GIVEN	INCH "	3 7 ¾ / 8	4 99 ¼	6 9 10	FIGURES GIVEN
10 31 ¼	12 34	SIDE CUT	" "	8 10 12	10 10 13	12 12 17	SIDE CUT OF HIP ON
16 40	12 43 ¼	OF JACKS	" "	15 10 16	16 9 15	18 10 18	VALLEY RAFTER

1\|0   9   8   7   6

Figure 9-34   A framing square showing readings for side cut of jack corresponding to 8-inch rise per foot run.

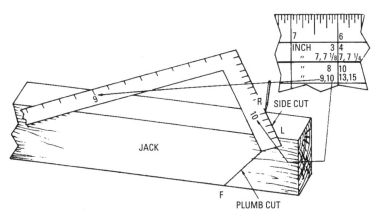

	7	6
INCH	3	4
"	7, 7 ⅛	7, 7 ¼
"	8	10
"	9,10	13,15

JACK · R · SIDE CUT · L · F · PLUMB CUT · 9 · 10

Figure 9-35   Method of placing a framing square on jack to lay off side cut for an 8-inch rise.

and 12 on the body will give the side cut at the ridge when there is no pitch (at *M*), as shown in Figure 9-37. Placing the square on the jack with the same setting numbers (12, 12) as at *S*, will give the face cut for the jack when framed to a 45° hip with zero pitch (that is, when all the timbers lie in the same plane).

Octagon Rafters

On an octagon (or eight-sided) roof, the rafters joining the corners are called *octagon rafters* and are a little longer than the common rafter and shorter than the hip or valley rafters of a square building of the same span. Figure 9-38 shows the relation between the run

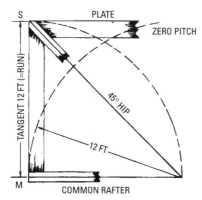

Figure 9-36 A roof with zero pitch shows the common rafter and the tangent as being the same length.

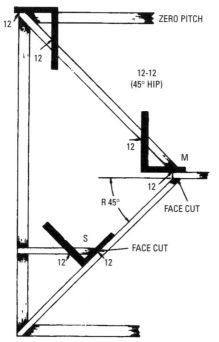

Figure 9-37 Zero-pitch square 45° roof shows application of the framing square to give side cuts at the ridge.

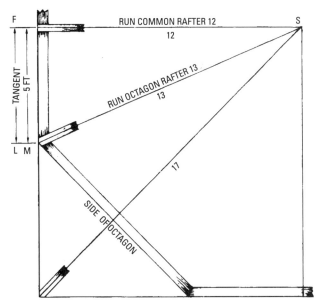

Figure 9-38 **Details of an octagon roof showing relation in length between common and octagon rafters.**

of an octagon and a common rafter as being as 13 is to 12. That is, for each foot run of a common rafter, an octagon rafter would have a run of 13 inches. Hence, to lay off the top or bottom cut of an octagon rafter, place the square on the timber with the 13 on the tongue and the rise of the common rafter per foot run on the blade (see Figure 9-39). Figure 9-40 shows the method of laying out the top and bottom cut with the 13-rise setting.

The length of an octagon rafter may be obtained by scaling the diagonal on the square for 13 on the tongue and the rise in inches per foot run of a common rafter, and multiplying by the number of feet run of a common rafter. The principle involved in determining the amount of backing of an octagon rafter (or any other polygon) is the same as for hip rafters. The backing is determined by the tangent of the angle whose adjacent side is one-half the rafter thickness and whose angle is equivalent to one-half the center angle.

Trusses

There are definite savings in material and labor requirements with preassembled wood roof trusses. They make truss framing an

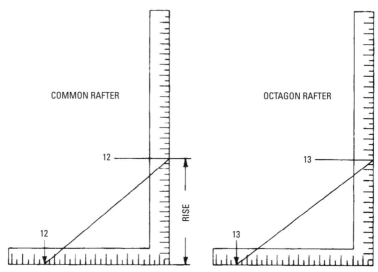

Figure 9-39   For equal rise, the run of octagon rafters is 13 inches, to 12 inches for the common rafters.

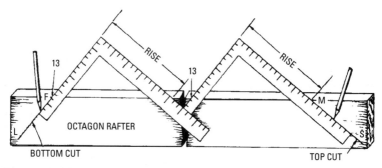

Figure 9-40   Method of laying-off bottom and top cuts of an octagon rafter with a square using the 13 rise setting.

effective means of cost reduction in small-dwelling construction (see Figure 9-41). In a 26-foot × 32-foot dwelling, for example, the use of trusses can result in a substantial cost savings and a reduction in use of lumber of almost 30 percent as compared with conventional rafter-and-joist construction. In addition to cost savings, roof trusses offer other advantages of increased flexibility for interior planning, and added speed and efficiency in site erection procedures. Today, most homes built in the United States use trusses. Not only may the framing lumber be smaller in dimension than in conventional framing, but trusses may also be spaced 24 inches on center as compared to the usual 16-inch spacing of rafter and joist construction.

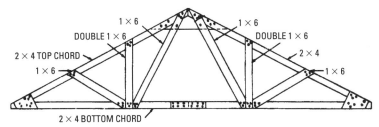

Figure 9-41   Wood roof truss for small dwellings.

Trusses are built of stress-rated lumber fastened together with gang nail plates. The clear span of truss construction permits use of nonbearing partitions so that it is possible to eliminate the extra top plate required for bearing partitions used with conventional framing. It also permits a small floor girder to be used for floor construction since the floor does not have to support the bearing partition and help carry the roof load.

Aside from direct benefits of reduced cost and savings in material and labor requirements, roof trusses offer special advantages in helping to speed up site erection and overcome delays caused by weather conditions. These advantages are reflected not only in improved construction methods but also in further reductions in cost. With preassembled trusses, a roof can be put over the job quickly to provide protection against the weather. Trusses are made and delivered to the job site by specialty manufacturers who also provide design services.

Dormers

The term *dormer* is given to any window protruding from a roof. The general purpose of a dormer may be to provide light or to add to the architectural effect.

In general construction, the following are three types of dormers:

- Dormers with flat-sloping roofs, but with less slope than the roof in which they are located (see Figure 9-42)

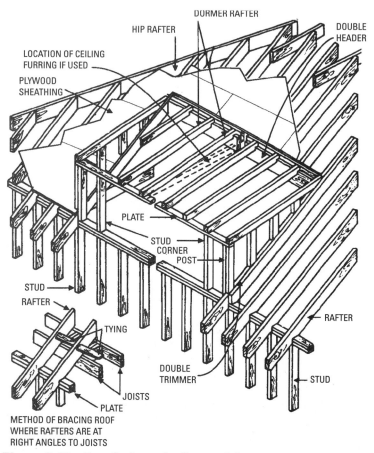

Figure 9-42 Detail view of a flat-roof dormer.

- Dormers with roofs of the gable type at right angles to the roof
- A combination of these types, which gives the hip-type dormer (see Figure 9-43)

When framing the roof for a dormer window, an opening is provided in which the dormer is later built. As the spans are usually short, light material may be used.

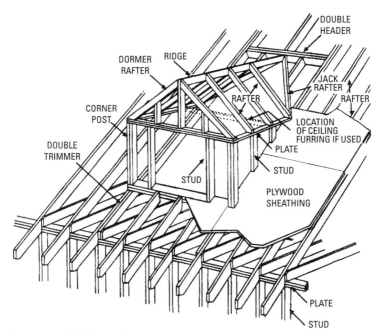

Figure 9-43 Detail view of a hip-roof dormer.

Summary

There are numerous forms of roofs and an endless variety of roof shapes. The frames of most roofs are made up of timbers called rafters. The terms used in connection with roofs are span, total rise, total run, unit of run, rise in inches, pitch, cut of roof, line length, and plumb and level lines.

Rafters are the supports for the roof covering and serve in the same manner, as do joists for floors or do studs for the walls. Rafters are constructed from ordinary 2-inch × 6-inch, 2 × 8, or 2 × 10 lumber, spaced 16 to 24 inches on center. Various kinds of rafters used in roof construction are common, hip, valley, jack, and octagon. The length of a rafter may be found in several ways (by calculation, with a steel framing square, or with the aid of a framing table).

Definite savings in material and labor with preassembled wood roof trusses make truss framing an effective means of cost reduction in small buildings. In addition to cost savings, roof trusses offer advantages of increased flexibility for interior planning, and added speed and efficiency in site-erection procedures. Light-wood trusses have been developed that permit substantial savings because they may be spaced 24 inches on center, as compared to the usual 16-inch spacing.

Review Questions

1. Name the various kinds of rafters used in roof construction.
2. Name the various terms used in connection with roofs.
3. What is preassembled truss roofing?
4. What are some advantages of preassembled truss roofing?
5. What are jack rafters and octagon rafters?
6. How can the length of a rafter be found?
7. Where would an octagon rafter be used?
8. List three common rafter tails.
9. Why would you have need for a flush hip rafter miter cut?
10. Where is the bird's eye on a rafter?
11. Jack rafters are virtually _____ common rafters.
12. Why is the term cripple applied to a rafter?

Chapter 10

Framing Chimneys and Fireplaces

Although the carpenter is ordinarily not concerned with the building of the chimney, it is necessary to be acquainted with the methods of framing around it (see Figure 10-1). Many contemporary homes have pre-fabricated fireplaces installed with their own set of instructions included for the carpenter and the crew.

The following minimum requirements are recommended:

- No wooden beams, joists, or rafters should be placed within 2 inches of the outside face of a chimney. No woodwork should be placed within 4 inches of the back wall of any fireplace.

- No studs, furring, lathing, or plugging should be placed against any chimney or in the joints thereof. Wooden construction should be set away from the chimney, or the plastering should be directly on the masonry, on metal lathing, or on non-combustible furring material.

- The walls of fireplaces should never be less than 8 inches thick if brick, or 12 inches if stone.

Prefabricated Fireplaces

A minimum amount of framing is necessary for prefabricated fireplaces (see Figure 10-2). They are placed on the slab or on the pre-poured spot in the case of a basement in the house. Some use brick or tile on the floor underneath the fireplace wood-burning section. Wood, of course, must be kept at a distance from the chimney's path up to the roof and into its enclosure on the roof.

The fireplace being installed in Figure 10-3 is sitting on concrete blocks and will have a gas-fed piece of pipe with holes along the top to furnish an ignition for logs laid on top of it. This one came with firebrick inside the fire chamber. These have dampers that are easily operated. The damper can control or stop the escape of room-heat when not in use.

Contemporary Design

Figure 10-4 shows an example of a very contemporary fireplace. This type is one that is usually installed after the house has been built and

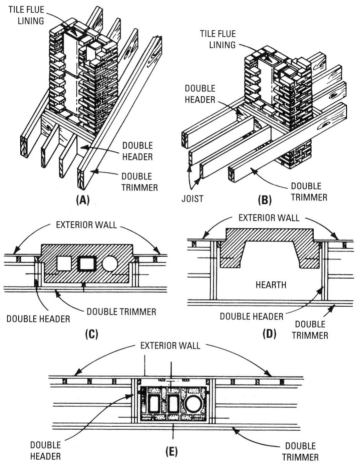

Figure 10-1   Framing around chimneys and fireplaces: (A) roof framing around chimney; **(B)** floor framing around chimney; **(C)** framing around chimney above fireplace; **(D)** floor framing around fireplace; **(E)** framing around concealed chimney above fireplace.

lived in for some time. The rocks (encased in a box) provide some protection from flying embers. The damper can be easily controlled by the butterfly knob at the junction of the fireplace with the exhaust tube. In this case, the fireplace is more a piece of furniture than a functional source of heat.

Figure 10-2   Framed-in prefab fireplace sitting on bricks. Note the double exhaust pipe where once the house would have had a solid masonry construction. This type is primarily for decoration. However, it can be used for heat if necessary.

Figure 10-3   Prefab fireplace sitting on concrete blocks.

Figure 10-4 A contemporary fireplace, usually installed after a house is built.

Summary

The minimum requirements recommended for fireplaces include the following:

- No wooden beams, joists, or rafters should be placed within 2 inches of the outside face of a chimney.
- No studs, furring, lathing, or plugging should be placed against any chimney or in the joints thereof.
- The walls of fireplaces should never be less than 8 inches thick if brick, or 12 inches if stone.

A minimum amount of framing is necessary for prefabricated fireplaces. They are placed on the slab or on the prepoured spot in the case of a basement in house. Some prefabricated fireplaces use brick or tile on the floor underneath the fireplace's wood-burning section.

Many freestanding fireplaces (in contemporary surroundings) are nothing more than a piece of furniture and are chosen for their ability to contribute to the decor of the room.

The floor around a contemporary fireplace may be covered with small rocks to protect from flying sparks. Of course, this isn't necessary if the floor is tiled, or is made of brick or concrete.

Review Questions

1. What type of fireplace is used today in modern house-building?

2. How far should wooden beams, joists, or rafters be kept from the outside face of a chimney?

3. The walls of fireplaces should never be less than _____ inches thick if brick, or 12 inches if stone.

4. Why are double trimmers or double headers needed around a fireplace?

5. Where is the hearth located in a fireplace installation?

Chapter 11

Roofs and Roofing

A roof includes the roof cover (the outer layer that protects against rain, snow, and wind), the sheathing to which it is fastened, and the framing (rafters) that supports the whole structure.

The term *roofing* (or *roof cover*) refers to the outermost part of a roof. Because of its exposure, roofing usually has a relatively limited life. It is made to be readily replaceable. It may be made of many different materials, including the following:

- *Wood*—These are usually in the form of shingles that are uniform, machine-cut; or hand-cut shakes (see Figure 11-1).

Figure 11-1 Wood shakes handsomely top this lovely home. Wood shingles, which give a uniform appearance, are also available. *(Courtesy of Scholz Homes, Inc.)*

- *Metal or aluminum*—This simulates other kinds of roofing.
- *Slate*—This may be the natural product, or rigid manufactured slabs, often of cement-asbestos.
- *Tile*—This is a burned clay or shale product. Several standard types are available.
- *Built-up covers of asphalt- or tar-impregnated felts*—These may have *moppings* of hot tar or asphalt between the plies

and a mopping of tar or asphalt overall. However, with tar-felt roofs, the top is usually covered with embedded gravel or crushed slag.

- *Roll roofing*—As the name implies, this is marketed in rolls containing approximately 108 square feet. Each roll is usually 36 inches wide and may be plain or have a coating of colored mineral granules. The base is a heavy asphalt impregnated felt.

- *Asphalt shingles*—These are usually in the form of strips with two, three, or four tabs per unit. These shingles are asphalt, with the surface exposed to the weather heavily coated with mineral granules. Because of their fire-resistance, cost, and reasonably good durability, this is the most popular roofing material for residences (see Figure 11-2). Asphalt shingles are available in a wide range of colors, including back and white.

Figure 11-2 Asphalt shingles being installed on a wood-shingle roof. Asphalt is now the most popular roofing.

- *Glass-fiber shingles*—These are made partly of a glass-fiber mat that is waterproof and partly of asphalt. Like asphalt shingles, glass-fiber shingles come with self-sealing tabs and carry a Class-A fire-resistance warranty (see Figure 11-3). For the do-it-yourselfer, they may be of special interest because they are lightweight, about 220 pounds per 100 square feet (see Figure 11-4).

Figure 11-3 Glass-fiber shingles are light in weight and have a high fire-resistance rating. *(Courtesy of Owens-Corning)*

Figure 11-4 These shingles look like slate, but they are actually glass fiber. *(Courtesy of Owens-Corning)*

Slope of Roofs

The slope of the roof is a factor in the choice of roofing materials and in the method used to put them in place. The lower the pitch of the roof, the greater is the chance of wind and water getting under the shingles. Interlocking cedar shingles resist this wind prying better than the standard asphalt shingles. For roofs with less than a 4-inch slope per foot, do not use standard asphalt. Roll roofing can be used with pitches down to 2 inches. For very low-pitched slopes, use built-up roofing.

Aluminum strip roofing virtually eliminates the problem of wind prying, but it is noisy. Most homeowners object to the noise during a rainstorm. Even on porches, this noise is often annoying inside the house.

Spaced roofing boards are sometimes used with cedar shingles as an economy measure and to allow an air space for ventilation.

For drainage, most roofs should have a certain amount of slope. Roofs with tar-and-gravel coverings are theoretically satisfactory when built level, but standing water may ultimately do harm. If you can avoid a flat roof, do so. Level roofs are common on industrial and commercial buildings.

Selecting Roofing Materials

Roofing materials are commonly sold by dealers or manufacturers based on quantities sufficient to cover 100 square feet. This quantity is commonly termed *one square*. When ordering roofing material, it will be well to make allowance for waste (such as in hips, valleys, and starter courses). This applies in general to all types of roofing.

The slope of the roof and the strength of the framing are the first determining factors in choosing a suitable covering. If the slope is slight, there will be danger of leaks with a wrong kind of covering, and excessive weight may cause sagging, which is unsightly and adds to the difficulty of keeping the roof in repair. The cost of roofing depends largely on the type of roof to be covered. A roof having ridges, valleys, dormers, or chimneys will cost considerably more to cover than one having a plain surface. Very steep roofs are also more expensive than those with a flatter slope, but most roofing materials last longer on steep grades than on low-pitched roofs. Frequently, nearness to supply centers permits the use (at lower cost) of the more durable materials instead of the commonly lower-priced, shorter-lived ones.

In considering cost, you should keep in mind maintenance, repair, and the length of service expected from the building. A permanent structure warrants a good roof, even through the first cost is somewhat high. When the cost of applying the covering is high in comparison with the cost of the material, or when access to the roof is hazardous, the use of long-lived material is warranted. Unless insulation is required, semi-permanent buildings and sheds are often covered with low-grade roofing.

Frequently, the importance of fire resistance is not recognized, and at other times it is wrongly stressed. It is essential to have a covering that will not readily ignite from glowing embers. The building regulations of many cities prohibit the use of certain types of roofing in congested areas where fires may spread rapidly. Underwriters Laboratories, Inc., has grouped many different kinds and brands of roofing in classes from A to C according to the protection offered against spread of fire. Class A is the best.

The appearance of a building can be changed materially by using the various coverings in different ways. Wood shingles and slate are often used to produce architectural effects. The roofs of buildings in a farm group should harmonize in color, even though similarity in contour is not always feasible.

All coal-tar pitch roofs should be covered with a mineral coating, because when fully exposed to the sun, they deteriorate. Observation has shown that, in general, roofing materials with light-colored surfaces absorb less heat than those with dark surfaces. Considerable attention should be given to the comfort derived from a properly insulated roof. A thin uninsulated roof gives the interior little protection from heat in summer and cold in winter. Discomfort from summer heat can be lessened to some extent by ventilating the space under the roof. None of the usual roof coverings have any appreciable insulating value. If it is necessary to reroof, consideration should be given to the feasibility of installing extra insulation under the roofing.

Roll Roofing

Roll roofing (see Figure 11-5) is an economical cover especially suited for roofs with low pitches. It also is sometimes used for valley flashing instead of metal. Roll roofing has a base of heavy, asphalt-impregnated felt with additional coatings of asphalt, which are dusted to prevent adhesion in the roll. The weather surface may be plain or covered with fine mineral granules. Many different colors are available. One edge of the sheet is left plain (no granules) where the lap cement is applied. For best results, the sheathing must be tight (preferably 1 × 6 tongue-and-groove, or plywood). If the sheathing is smooth, with no cupped boards or other protuberance,

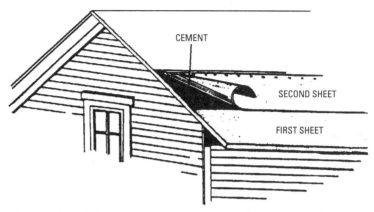

Figure 11-5   First and second strips of roll roofing installed.

the slate-surfaced roll roofing will withstand a surprising amount of abrasion from foot traffic, although it is not generally recommended for that purpose. Windstorms are the most relentless enemy of roll roofing. If the wind gets under a loose edge, almost certainly a section will be blown off.

The Built-Up Roof

A built-up roof is constructed of sheathing paper, a bonded base sheet, perforated felt, asphalt, and surface aggregates (see Figure 11-6). The sheathing paper comes in 36-inch-wide rolls and has approximately 500 square feet per roll. It is a rosin-size paper and is used to prevent asphalt leakage to the wood deck. The base sheet

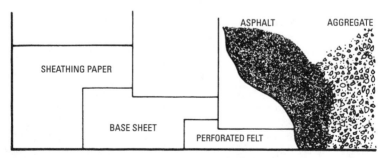

Figure 11-6   Sectional plan of a built-up roof.

is a heavy, asphalt-saturated felt that is placed over the sheathing paper. It is available in 1-, 1$^1/_2$-, and 2-square rolls. The perforated felt is one of the primary parts of a built-up roof. It is saturated with asphalt and has tiny perforations throughout the sheet. The perforations prevent air entrapment between the layers of felt. The perforated felt is 36 inches wide and weighs approximately 15 pounds per square. Asphalt is also one of the basic ingredients of a built-up roof. There are many different grades of asphalt, but the most common are:

- Low melt
- Medium melt
- High melt
- Extra-high melt

Prior to the application of the built-up roof, the deck should be inspected for soundness. Wood board decks should be constructed of $^3/_4$-inch seasoned lumber. Any knotholes larger than 1 inch should be covered with sheet metal. If plywood is used as a roof deck, it should be placed at right angles to the rafters and be at least $^1/_2$ inch in thickness.

The first step in the application of a built-up roof is the placing of sheathing paper and base sheet. The sheathing paper should be lapped 2 inches and secured with just enough nails to hold it in place. The base sheet is then placed with 2-inch side laps and 6-inch end laps. The base sheet should be secured with $^1/_2$-inch-diameter–head galvanized roofing nails placed 12 inches on center on the exposed lap. Nails should also be placed down the center of the base sheet. The nails should be placed in two parallel rows 12 inches apart.

Each sheet is then coated with a uniform layer of hot asphalt. While the asphalt is still hot, layers of roofing felt are placed. Each sheet should be lapped 19 inches, leaving an exposed lap of 17 inches.

Once the roofing felt is placed, a gravel stop is installed around the deck perimeter (see Figure 11-7). Two coated layers of felt should extend 6 inches past the roof decking where the gravel stop is to be installed. When the other plies are placed, the first two layers are folded over the other layers and mopped in place. The gravel stop is then placed in a $^1/_8$-inch-thick bed of flashing cement and securely nailed every 6 inches. The ends of the gravel stop should be lapped 6 inches and packed in flashing cement.

After the gravel stop is placed, the roof is flooded with hot asphalt and the surface aggregate is embedded in the flood coat. The

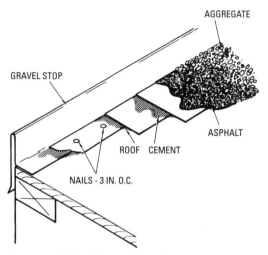

Figure 11-7 Illustrating the gravel stop.

aggregates should be hard, dry, opaque, and free of any dust or foreign matter. The size of the aggregates should range from $1/4$ inch to $5/8$ inch. When the aggregate is piled on the roof, it should be placed on a spot that has been mopped with asphalt. This technique ensures proper adhesion in all areas of the roof.

Wood Shingles

The better grades of wood shingles are made of cypress, cedar, and redwood and are available in lengths of 16 and 18 inches and thicknesses at the butt of $5/16$ inch and $7/16$ inch, respectively. They are packaged in bundles of approximately 200 shingles in random widths from 3 to 12 inches.

An important requirement in applying wood shingles is that each shingle should lap over the two courses below it, so that there will always be at least three layers of shingles at every point on the roof. This requires that the amount of shingle exposed to the weather (the spacing of the courses) should be less than one-third the length of the shingle. Thus in Figure 11-8, $5^{1}/_{2}$ inches is the maximum amount that 18-inch shingles can be laid to the weather and have an adequate amount of lap. This is further shown in Figure 11-9A.

If the shingles are laid more than one-third of their length to the weather, there will be a space, as shown by *MS* in Figure 11-9B, where only two layers of shingles will cover the roof. This is

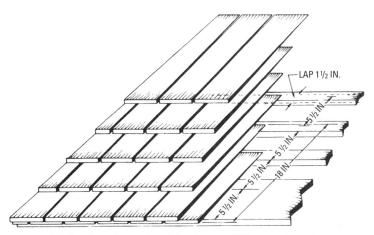

Figure 11-8 **Section of a shingle roof illustrating the amount of shingle that may be exposed to the weather, as governed by the lap.**

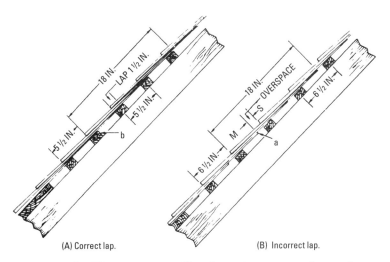

(A) Correct lap.

(B) Incorrect lap.

Figure 11-9 **The amount of lap is an important factor in applying wood shingles.**

objectionable, because if the top shingle splits above the edge of the shingle below, water will leak through. The maximum spacing to the weather for 16-inch shingles should be $4^{7}/_{8}$ inches, and for 18-inch shingles should be $5^{1}/_{2}$ inches. Strictly speaking, the amount of lap should be governed by the pitch of the roof. The maximum spacing may be followed for roofs of moderate pitch, but for roofs of small pitch, more lap should be allowed, and for a steep pitch the lap may be reduced somewhat, but it is not advisable to do so. Wood shingles should not be used on pitches less than 4 inches per foot.

Table 11-1 shows the number of square feet that 1000 shingles (five bundles) will cover for various exposures. This table does not allow for waste on hip and valley roofs.

Table 11-1   Space Covered by 1000 Shingles

Exposure to Weather, in inches	Area Covered, in Sq. Ft
$4^{1}/_{4}$	118
$4^{1}/_{2}$	125
$4^{3}/_{4}$	131
5	138
$5^{1}/_{2}$	152
6	166

Shingles should not be laid too close together, for they will swell when wet, causing them to bulge and split. Seasoned shingles should not be placed with their edges nearer than $^{3}/_{16}$ inches when laid. It is advisable to soak the bundles thoroughly before opening.

Great care must be used in nailing wide shingles. When they are more than 8 inches in width, they should be split and laid as two shingles. The nails should be spaced such that the distance between them is as small as is practical; thus directing the contraction and expansion of the shingle toward the edges. This lessens the danger of wide shingles splitting in or near the center and over joints beneath. Shingling is always started from the bottom and laid from the eaves or cornice up.

There are various methods of laying shingles, the most common known as:

- The straightedge
- The chalk-line
- The gage-and-hatchet

The *straightedge method* is one of the oldest. A straightedge having a width equal to the spacing to the weather or the distance between courses is used. This eliminates measuring, it being necessary only to keep the lower edge flush with the lower edge of the course of shingles just laid. The upper edge of the straightedge is then in line for the next course. This is considered the slowest of the three methods.

The *chalk-line method* consists of snapping a chalk line for each course. To save time, two or three lines may be snapped at the same time, making it possible to carry two or three courses at once. It is faster than the straightedge method, but not as fast as the gage-and-hatchet method.

The *gage-and-hatchet method* is extensively used in western states. The hatchet used is either a lathing or a box-maker's hatchet (see Figure 11-10). Figure 11-11 shows hatchet gages used to measure the space between courses. The gage is set on the blade at a distance from the hatchet poll equal to the exposure desired for the shingles.

Nail as close to the butts as possible if the nails will be well covered by the next course. Only galvanized shingle nails should be used. The third shingle nail is slightly larger in diameter than the third common nail, and has a slightly larger head.

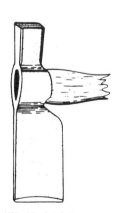

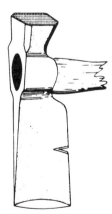

(A) Lathing hatchet.   (B) Box-maker's hatchet.

Figure 11-10   Hatchets used in shingling.

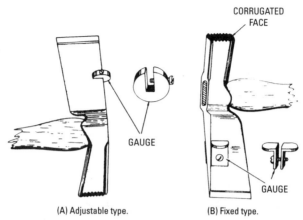

(A) Adjustable type.   (B) Fixed type.

Figure 11-11   Shingling hatchets and gages.

Hips

The hip is less likely to leak than any other part of the roof, because the water runs away from it. However, since it is so prominent, the work should be well done. Figure 11-12 shows the method of cutting shingle butts for a hip roof. After courses 1 and 2 are laid, the top corners over the hip are trimmed off with a sharp shingling hatchet kept keen for that purpose. Shingle 3 is trimmed with a butt cut so as to continue the straight line of courses, and again on dotted line 4, so that shingle A, of the second course, squares against it. This process is repeated from side to side, each alternately lapping at the hip joint. When gables are shingled, this same method may be used up the rake of the roof if the pitch is moderate to steep. It cannot be effectively used with flat pitches. The shingles used should be ripped to uniform width.

For best construction, metal shingles should be laid under the hip shingles (see Figure 11-13). These metal shingles should correspond in shape to that of the hip shingles. They should be at least 7 inches wide and large enough to reach well under the metal shingles of the course above, as at w. At a, the metal shingles are laid so that the lower end will just be covered by the hip shingle of the course above.

Valleys

In shingling a valley, first a strip of sheet metal or roll roofing (ordinarily 20 inches wide) is laid in the valley. Figure 11-14 illustrates an open-type valley. Here the dotted lines show the aluminum or other

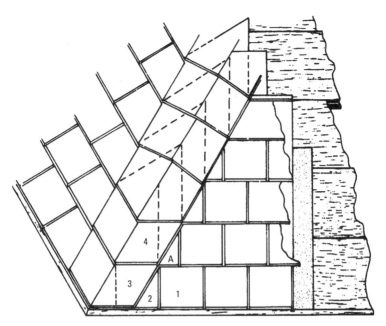

Figure 11-12   Hip-roof shingling.

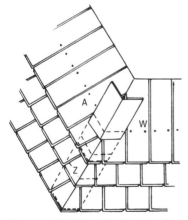

Figure 11-13   Method of installing metal shingles under wooden shingles.

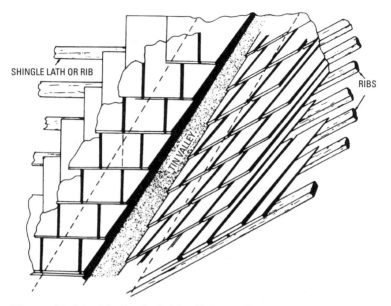

SHINGLE LATH OR RIB

TIN VALLEY

RIBS

Figure 11-14   Method of shingling a valley.

material used as flashing under the shingles. If the pitch is above 30°, then a width of 16 inches is sufficient; if flatter, the width should be more. In a long valley, its width between shingles should increase in width from 1 inch at the top to 2 inches at the bottom. This is to prevent ice or other objects from wedging when slipping down. The shingles taper to the butt (the reverse of the hip) and need no reinforcing, because the thin edge is held and protected from splitting off by the shingle above it. Care must always be taken to nail the shingle nearest the valley as far from it as practical by placing the nail higher up.

Asphalt Shingles

Asphalt shingles are made in strips of two, three, or four units or tabs joined together, as well as in the form of individual shingles. When laid, strip shingles furnish practically the same pattern as individual shingles. Both strip and individual types are available in different shapes, sizes, and colors to suit various requirements.

Asphalt shingles must be applied on slopes having an incline of 4 inches or more to the foot. Before the shingles are laid, the underlayment should be placed. The underlayment should be 15-pound asphalt-saturated felt. This material should be placed with

2-inch side laps and 4-inch end laps (see Figure 11-15). The underlayment serves two purposes:

- It acts as a secondary barrier against moisture penetration.
- It acts as a buffer between the resinous areas of the decking and the asphalt shingles.

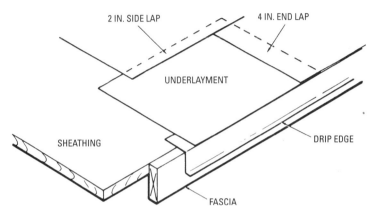

Figure 11-15   Application of the underlayment.

A heavy felt should not be used as underlayment. The heavy felt would act as a vapor barrier and would permit the accumulation of moisture between the underlayment and the roof deck.

The roof deck may be constructed of well-seasoned 1-inch × 6-inch tongue-and-groove sheathing or plywood. The boards should be secured with two 8d nails in each rafter. Plywood should be placed with the long dimension perpendicular to the rafters. The plywood should never be less than $5/8$ inch thick.

To efficiently shed water at the roof's edge, a drip edge is usually installed. A drip edge is constructed of corrosion-resistant sheet metal. It extends 3 inches back from the roof edge. To form the drip-edge the sheet metal is bent down over the roof edges.

The nails used to apply asphalt shingles should be hot galvanized nails with large heads, sharp points, and barbed shanks. The nails should be long enough to penetrate the roof decking at least $3/4$ inch.

To ensure proper shingle alignment, horizontal and vertical chalk lines should be snapped on the underlayment. It is usually recommended that the lines be placed 10 or 20 inches apart. The first course of shingles placed is the starter course. This is used to back up the first regular course of shingles and to fill in the spaces between the tabs. It is placed with the tabs facing up the roof and is allowed to project 1 inch over the rake and eave (see Figure 11-16). To ensure that all cutouts are covered, 3 inches should be cut off the first starter shingle.

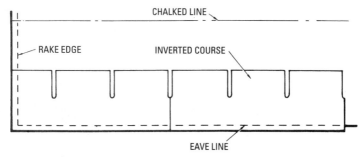

Figure 11-16   The starter course.

Once the starter course has been placed, the different courses of shingles can be laid. The first regular course of shingles should be started with a full shingle; the second course with a full shingle, minus one-half a tab; the third course is started with a full shingle (see Figure 11-17); and the process is repeated. As the shingles are placed, they should be properly nailed (see Figure 11-18). If a three-tab shingle is used, a minimum of four nails per strip should be used. The nails should be placed $5\frac{5}{8}$ inches from the bottom of the shingle and should be located over the cutouts. The nails on each end of the shingle should be located one inch from the end. The nails should be driven straight and flush with the surface of the shingle.

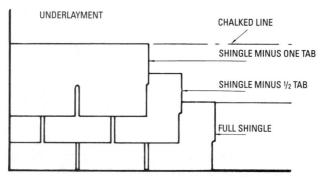

Figure 11-17   Application of the starter shingles.

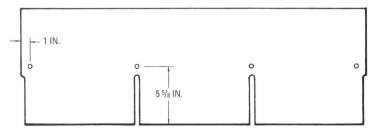

Figure 11-18   The proper placement of nails.

Figure 11-19 shows two roofers bringing up the air-hose to drive their stapler in anchoring the shingles to the sheathing. Figure 11-20 shows the mess that can accumulate if the roofers do not properly dispose of the wrappings of the shingles. This type of environment can become a safety hazard very quickly. If the wind starts to build up speed, the paper will create a public-relations nightmare for the builder as the neighbors start complaining.

If there is a valley in the roof, it must be properly flashed. The two materials that are most often used for valley flashing are 90-pound mineral-surfaced asphalt roll roofing or sheet metal. The flashing

Figure 11-19   Roofers pulling up their air hose to continue up-ward in the application of shingles.

Figure 11-20 Shingle wrappings create a safety hazard and become unsightly.

is 18 inches in width. It should extend the full length of the valley. Before the shingles are laid to the valley, chalked lines are placed along the valley. The chalk lines should be 6 inches apart at the top of the valley and should widen $1/8$ inch per foot as they approach the eave line. The shingles are laid up to the chalked lines and trimmed to fit.

Hips and ridges are finished by using manufactured hip and ridge units, or hip and ridge units cut from a strip shingle. If the unit is

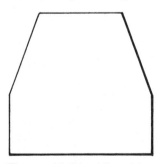

Figure 11-21 Hip shingle.

cut from a strip shingle, the two cut lines should be cut at an angle (see Figure 11-21). This will prevent the projection of the shingle past the overlaid shingle. Each shingle should be bent down the center so that there is an equal distance on each side. In cold weather, the shingles should be warmed before they are bent. Starting at the bottom of the hip or at the end of a ridge, the shingles are placed with a 5-inch exposure. To secure the shingles, a nail is placed on each side of the shingle. The nails should be placed $5\,{}^1\!/_2$ inches back from the exposed edge and 1 inch up from the side.

If the roof slope is particularly steep (specifically if it exceeds $60°$ or 21 inches per foot), then special procedures are required for securing the shingles (see Figure 11-22).

Two other details are worth noting. For neatness when installing asphalt shingles, the courses should meet in a line above any dormer (see Figure 11-23). In addition, of course, ventilation must be provided for an asphalt roof. All roofs should be ventilated.

Slate

Slate is not used as much as it once was, but it is still used. The process of manufacture is to split the quarried slate blocks horizontally to a suitable thickness, and to cut vertically to the approximate sizes required. The slates are then passed through planers and, after the operation, are ready to be reduced to exact dimensions on rubbing beds, or by air tools and other special machinery.

Roofing slate is usually available in various colors and in standard sizes suitable for the most exacting requirements. On all boarding to be covered with slate, asphalt-saturated rag felt of certain specified thickness is required. This felt should be laid in a horizontal layer with joints lapped toward the eaves and at the ends at least 2 inches. A well-secured lap at the end is necessary to hold the felt in place properly, and to protect the structure until covered by the slate. In laying the slate, the entire surface of all main and porch roofs should be covered with slate in a proper and watertight manner.

The slate should project 2 inches at the eaves and 1 inch at all gable ends, and should be laid in horizontal courses with the standard 3-inch head lap. Each course should break joints with the preceding one. Slates at the eaves or cornice line should be doubled and canted $\,{}^1\!/_4$ inch by a wooden cant strip. Slates overlapping sheet-metal work should have the nails so placed as to avoid puncturing the sheet metal. Exposed nails should be used only in courses where unavoidable. Neatly fit the slate around any pipes, ventilators, and so on.

Nails should not be driven in so far as to produce a strain on the slate. Cover all exposed nails heads with elastic cement. Hip slates and ridge slates should be laid in elastic cement spread thickly over unexposed surfaces. Build in, place all flashing pieces furnished by the sheeting contractor, and cooperate with him or her in doing the work of flashing. On completion, all slate must be sound, whole,

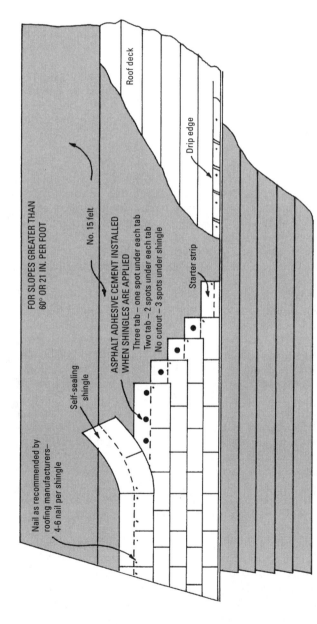

Figure 11-22 When a roof has a severe slope, special installation procedures are required for asphalt shingles.

Nail as recommended by roofing manufacturers— 4-6 nail per shingle

Self-sealing shingle

FOR SLOPES GREATER THAN 60° OR 21 IN. PER FOOT

No. 15 felt

ASPHALT ADHESIVE CEMENT INSTALLED WHEN SHINGLES ARE APPLIED

Three tab – one spot under each tab
Two tab – 2 spots under each tab
No cutout – 3 spots under shingle

Starter strip

Roof deck

Drip edge

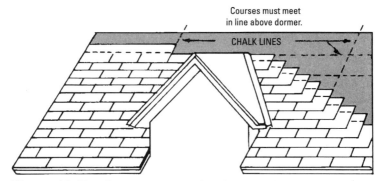

Figure 11-23 Arrangement of shingles when there is a dormer.

and clean, and the roof should be left in every respect tight and a neat example of workmanship.

Gutters and Downspouts

Most roofs require gutters and downspouts to carry the water away from the foundation (see Figure 11-24). They are made of

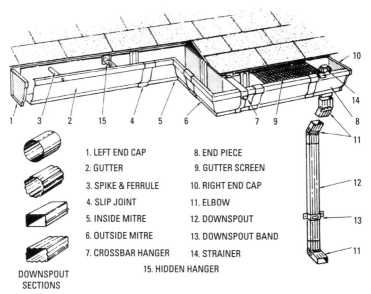

1. LEFT END CAP
2. GUTTER
3. SPIKE & FERRULE
4. SLIP JOINT
5. INSIDE MITRE
6. OUTSIDE MITRE
7. CROSSBAR HANGER

8. END PIECE
9. GUTTER SCREEN
10. RIGHT END CAP
11. ELBOW
12. DOWNSPOUT
13. DOWNSPOUT BAND
14. STRAINER

15. HIDDEN HANGER

DOWNSPOUT SECTIONS

Figure 11-24 Various metal gutter downspouts and fittings. *(Courtesy of Billy Penn Gutters)*

aluminum, steel, wood, or plastic. In regions of heavy snowfall, the outer edge of the gutter should be $\frac{1}{2}$ inch below the extended slope of the roof to prevent snow banking on the edge of the roof and causing leaks. The hanging gutter is best adapted to such construction.

Downspouts should be large enough to remove the water from the gutters. A common fault is to make the gutter outlet the same size as the downspouts. At 18 inches below the gutter, a downspout has nearly four times the water-carrying capacity of the inlet at the gutter. Therefore, a good-sized ending to the downspout should be provided. Wire baskets or guards should be placed at gutter outlets to prevent leaves and trash from collecting in the downspouts and causing damage during freezing weather.

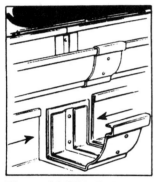

Figure 11-25 A gutter-section connector.

The most popular kind of gutter is made of aluminum. It comes in two common gages, 0.027 and 0.032, with the thicker material better, of course. Standard lengths are 10 feet, and they are joined by special connectors (see Figure 11-25) using either sheet-metal screws or blind rivets (blind rivets are simplest).

Any connection, however, represents an area that can leak. It is better to get so-called seamless gutter. Fabricators will custom-cut to fit. You will also save installation time. Seamless gutter is commonly 0.032-gage.

Gutters should have the proper slope for good runoff of water—about $\frac{1}{2}$ inch to every 10 feet (see Figure 11-26). Some people make the mistake of sloping one gutter according to the way the house

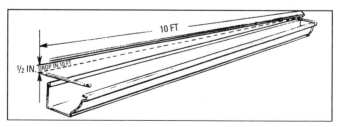

Figure 11-26 Gutter should slope $\frac{1}{2}$ inch per 10 feet.

appears. However, this can lead to errors because a house, although it may look level, never really is.

Summary

The roof of a building includes the roof cover (which is protection against rain, snow, and wind), the sheathing (which is a base for the roof cover), and the rafters (which are the support for the entire roof structure).

The term *roofing* refers to the outermost part of the roof. There are various types of roofing used, such as wood (which generally is in the form of shingles or shakes), aluminum, tile, roll roofing, asphalt, and glass fiber.

Various types and styles of flashing are used when a roof connects to any vertical wall (such as chimneys, outside walls, and so on). Flashing around chimneys and skylights is installed in the same general manner as for vertical walls. It is generally made from roll-roofing material, sheet metal, or aluminum bent to fit the contour of the vertical wall. It is essential for sealing joints.

Most roofs require rain gutters and downspouts to carry the water to the sewer or outlet. Gutters and downspouts are usually built of aluminum. Seamless 0.032 gutter is best. Downspouts should be large enough to remove the water from the gutters. Much gutter deterioration is caused by freezing water in low areas, rust, and restricted sections caused by leaves or other debris.

Review Questions

1. Name various types of roofing material.
2. What is flashing and why is it used?
3. Why can corrugated metal roofs be installed without roof sheathing?
4. How much coverage in square feet is one square of roofing?
5. What is a drip cap?
6. Why is the slope of the roof a factor in the choice of roofing material?
7. True or false, asphalt shingles come in strips.
8. Where is roll roofing acceptable for use?
9. What is a built-up roof?
10. The better grades of wood shingles are made of cypress, cedar, and _____.

Chapter 12

Skylights

A *skylight* is any window placed in the roof of a building or ceiling of a room for the admission of light and/or ventilation. Skylights are essential to lighting and ventilating top floors where roofs are flat (such as in factories) and where there is not much side lighting. In the home, they are often placed at the top of stairs or in a room where no side window is available.

Figure 12-1 shows a simple hinged skylight and detail of the hinge. The skylight may be operated from below by the control device, which has an adjustment eye in the support for securing the hinged sash at various degrees of opening. A skylight is often placed at the top of a flight of stairs leading to the roof, the projecting structure having framed in it the skylight and a doorway (see Figure 12-2).

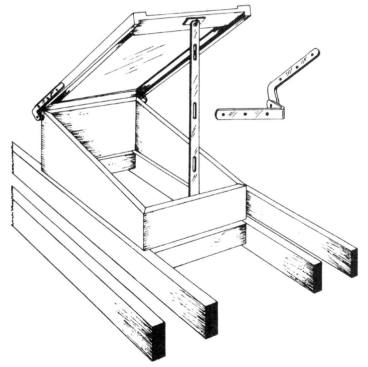

Figure 12-1   Hinged skylight framed into roof.

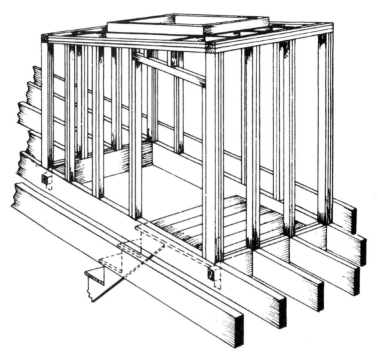

**Figure 12-2   View of entrance to roof, or projection framework
containing framed opening for skylight and doorway.**

Where fireproof construction is required, skylights are made of
metal. Figure 12-3 shows side-pivoted sash skylights. This is a type of
skylight desirable for engine and boiler rooms where a great amount
of steam and heat is generated. A storm coming in through them
would do little harm if the skylights were thoughtlessly left open. The
sash may be operated separately by pulleys, or all on one adjuster.
The ends may be stationary.

Wire glass should be used for factory skylights so that, if broken,
it will not fall and possibly injure someone below. Wire glass is cast
with the wire netting running through its center and is manufactured
in many styles and sizes.

Skylights are becoming more popular in residential dwellings (see
Figure 12-4). They come double insulated and can be opened for
ventilation. They are set in waterproof frames, which are permanently installed and sealed in the roof. A skylight can often help

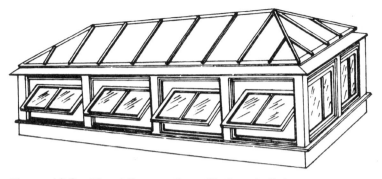

Figure 12-3   Metal fireproof ventilating skylight.

Figure 12-4   Skylight in a residence. It floods dark areas with light and can even help heat the home.

heat a house. A significant amount of radiant heat from the sun gets through.

Residential Skylights

The skylight shown in Figure 12-5 is positioned in this home to provide an openness and free-space look for the entrance, foyer, and family room. As can be seen, the makeshift ladder placed on

Figure 12-5 Skylight viewed from inside during house construction.

top of the scaffolding makes for a dangerous working condition when it is located over 25 feet off the floor. The octagonal design is incorporating (from a design standpoint) a square-domed skylight. This type of skylight does have the disadvantage of providing extra heat in the house during summer days. They are also noisy. The noise from traffic nearby can be both annoying and tiring. The skylight pictured does not open. The area from the roof to the ceiling, by way of the attic, must be finished off with drywall and painted.

The skylight shown in Figure 12-6 is a smaller type that serves well in a kitchen to make it a lighter and roomier place for cooking and serving food. This one is fixed. It requires a large light tunnel to be framed in and drywalled through the attic.

Figure 12-6 Smaller skylight used to brighten up a kitchen area.

Skylight Maintenance

Condensation may appear on the inner dome surface with sudden temperature changes or during periods of high humidity. These droplets are condensed moisture. Condensation will evaporate as conditions of temperature and humidity normalize.

Figure 12-7 shows how light-shaft installations can be used to present the light from the skylight to various parts of the room below. Figures 12-8 and 12-9 show how the original installations are made in houses under construction. Details and basic sizes are given, along with the roof pitch and slope chart. These will help you plan the installation from the start.

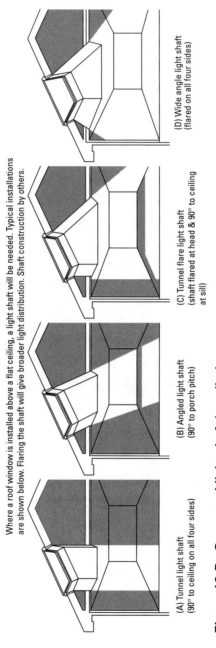

Where a roof window is installed above a flat ceiling, a light shaft will be needed. Typical installations are shown below. Flaring the shaft will give broader light distribution. Shaft construction by others.

(A) Tunnel light shaft
(90° to ceiling on all four sides)

(B) Angled light shaft
(90° to porch pitch)

(C) Tunnel flare light shaft
(shaft flared at head & 90° to ceiling at sill)

(D) Wide angle light shaft
(flared on all four sides)

Figure 12-7 Suggested light shaft installations. *(Courtesy of Andersen)*

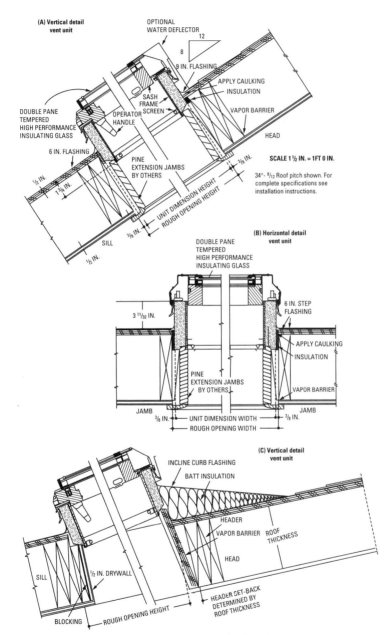

(A) Vertical detail vent unit

OPTIONAL WATER DEFLECTOR

12
8

9 IN. FLASHING

APPLY CAULKING

INSULATION

SASH FRAME SCREEN

VAPOR BARRIER

OPERATOR HANDLE

DOUBLE PANE TEMPERED HIGH PERFORMANCE INSULATING GLASS

HEAD

6 IN. FLASHING

PINE EXTENSION JAMBS BY OTHERS

SCALE 1 ½ IN. = 1FT 0 IN.

3/8 IN.

34°- 8/12 Roof pitch shown. For complete specifications see installation instructions.

½ IN.

1-3/4 IN.

UNIT DIMENSION HEIGHT
ROUGH OPENING HEIGHT

SILL

3/8 IN.

½ IN.

(B) Horizontal detail vent unit

DOUBLE PANE TEMPERED HIGH PERFORMANCE INSULATING GLASS

3 11/32 IN.

6 IN. STEP FLASHING

APPLY CAULKING

INSULATION

PINE EXTENSION JAMBS BY OTHERS

VAPOR BARRIER

JAMB

3/8 IN.

UNIT DIMENSION WIDTH

ROUGH OPENING WIDTH

JAMB

3/8 IN.

(C) Vertical detail vent unit

INCLINE CURB FLASHING

BATT INSULATION

HEADER

VAPOR BARRIER   ROOF THICKNESS

HEAD

SILL

½ IN. DRYWALL

HEADER SET-BACK DETERMINED BY ROOF THICKNESS

BLOCKING

ROUGH OPENING HEIGHT

Figure 12-8   Roof-window vent unit, in place. *(Courtesy of Andersen)*

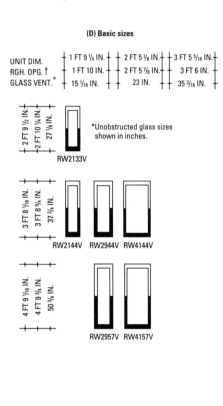

(D) Basic sizes

UNIT DIM.	1 FT 9¼ IN.	2 FT 5⅛ IN.	3 FT 5 5/16 IN.
RGH. OPG. †	1 FT 10 IN.	2 FT 5⅞ IN.	3 FT 6 IN.
GLASS VENT.*	15 1/16 IN.	23 IN.	35 3/16 IN.

*Unobstructed glass sizes shown in inches.

(E) Incline curb flashing rough openings

† WHEN INSTALLING UNITS WITH INCLINE CURB FLASHING USE THESE ROUGH OPENINGS.

UNIT	WIDTH DIM. A	HEIGHT DIM. B
2133	21 5/8 IN.	34 3/4 IN.
2144	21 5/8 IN.	45 1/2 IN.
2944	29 1/2 IN.	45 1/2 IN.
2957	29 1/2 IN.	58 3/4 IN.
4144	41 3/4 IN.	45 1/2 IN.
4157	41 3/4 IN.	58 3/4 IN.
HEADER SET BACK		
ROOF THICKNESS	DIM. C	
6 1/2 IN.	13/16 IN.	
8 1/2 IN.	1 1/8 IN.	
10 1/2 IN.	1 1/2 IN.	
12 1/2 IN.	1 13/16 IN.	

(F) Roof pitch/slope chart

ROOF PITCH	ROOF SLOPE
2/12	9° 26'
3/12	14°
4/12	18° 26'
5/12	22° 37'
6/12	26° 34'
7/12	30° 15'
8/12	33° 41'
9/12	36° 52'
10/12	39° 48'
11/12	42° 30'
12/12	45°
14/12	49° 24'
20/12	59°
40/12	70°
68/12	80°

Incline curb flashing is recommended for roof installations less than 18 1/2° (4/12 pitch) to 9° (2/12 pitch) minimum.

3/12 roof pitch (14°) shown. For complete specifications see installation instructions.

Figure 12-8 (continued)

If the dome is made of plastic, the outer dome surface may be polished with paste wax for added protection from outdoor conditions. If it is made of glass, you may want to wash it before installation and then touch up the finger marks after it is in place. Roofing mastic can be removed with rubbing alcohol or lighter fluid. Avoid petroleum-based or abrasive cleaners, especially on clear plastic domes. Roof

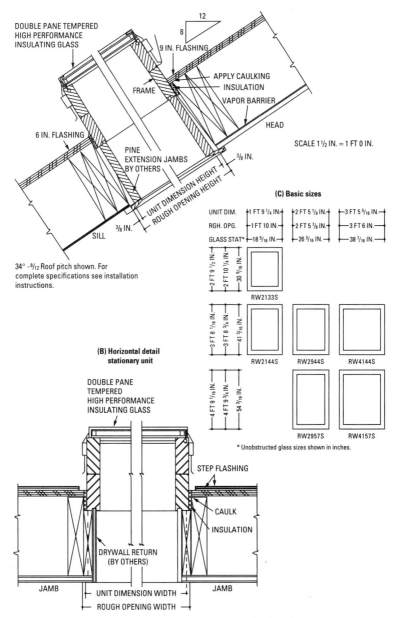

(A) Vertical detail stationary unit

DOUBLE PANE TEMPERED HIGH PERFORMANCE INSULATING GLASS

12
8

9 IN. FLASHING

FRAME

APPLY CAULKING
INSULATION
VAPOR BARRIER

HEAD

6 IN. FLASHING

PINE EXTENSION JAMBS BY OTHERS

3/8 IN.

UNIT DIMENSION HEIGHT

ROUGH OPENING HEIGHT

SCALE 1½ IN. = 1 FT 0 IN.

3/8 IN.

SILL

34° -8/12 Roof pitch shown. For complete specifications see installation instructions.

(B) Horizontal detail stationary unit

DOUBLE PANE TEMPERED HIGH PERFORMANCE INSULATING GLASS

STEP FLASHING

CAULK

INSULATION

DRYWALL RETURN (BY OTHERS)

JAMB

JAMB

UNIT DIMENSION WIDTH

ROUGH OPENING WIDTH

(C) Basic sizes

UNIT DIM.	1 FT 9 ¼ IN.	2 FT 5 ⅛ IN.	3 FT 5 5/16 IN.
RGH. OPG.	1 FT 10 IN.	2 FT 5 ⅞ IN.	3 FT 6 IN.
GLASS STAT*	18 5/16 IN.	26 3/16 IN.	38 7/16 IN.

2 FT 9 ½ IN.
2 FT 10 ¼ IN.
30 9/16 IN.

RW2133S

3 FT 8 1/16 IN.
3 FT 8 ¾ IN.
41 3/16 IN.

RW2144S   RW2944S   RW4144S

4 FT 9 1/16 IN.
4 FT 9 ¾ IN.
54 7/16 IN.

RW2957S   RW4157S

* Unobstructed glass sizes shown in inches.

Figure 12-9   Roof-window stationary unit, in place. *(Courtesy of Andersen)*

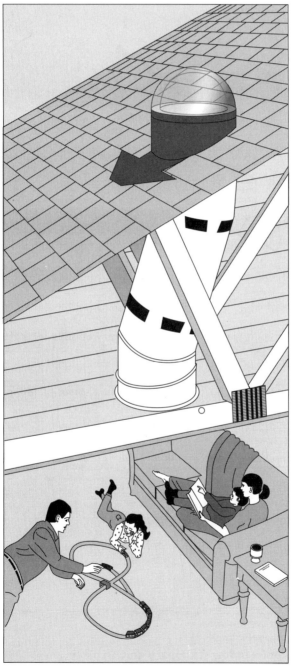

Figure 12-10 Skylight installation. *(Courtesy of ODL, Inc.)*

inspection should be conducted every two years to determine potential loosening of screws, cracked mastic, and other weather-related problems that may result from normal exposure to outdoor conditions.

Tube-Type Skylights

Newer tube-type skylights can be installed during house construction or added later. They are designed to provide maximum light throughput from a relatively small unit. They are right for areas where a larger, standard skylight may not be practical (see Figure 12-10).

Tube-type skylights come in a kit with everything needed, including illustrated instructions for the do-it-yourselfer. They install in a few hours with basic hand tools. There is no framing, drywalling, mudding, or painting required. They are available in both 10-inch and 14-inch diameters and therefore fit easily between 16-inch or 24-inch on-center rafters (see Figure 12-11).

Most people are concerned about skylights because they have heard of them leaking, especially during the winter with snow piling up, then melting. The skylight shown in Figure 12-12 has a one-piece roof flashing that eliminates leaks. Flashing is specific to the roof type and ensures a perfect fit. The 14-inch skylight spreads light up to 300 square feet. There is also an electric light kit available that makes the skylight into a standard light fixture at night and during dark periods of the day. It is designed to work from a wall switch and is a UL-approved installation (see Figure 12-13).

Installation

To install a tube-type skylight, follow these steps:

1. Locate the diffuser position on the ceiling.
2. Check the attic for any obstructions or wiring.
3. Locate the position on the roof for flashing and dome. If the skylight is being installed in new construction, you can make sure plumbing and electrical take the skylight into consideration during the construction phase.
4. Measure and cut an opening in the roof.
5. Loosen shingles and install the flashing. In new construction, it may be best to install the flashing before shingles are in place.
6. Insert the adjustable tube.
7. Attach the dome (see Figure 12-14).
8. Measure and cut an opening in the ceiling.

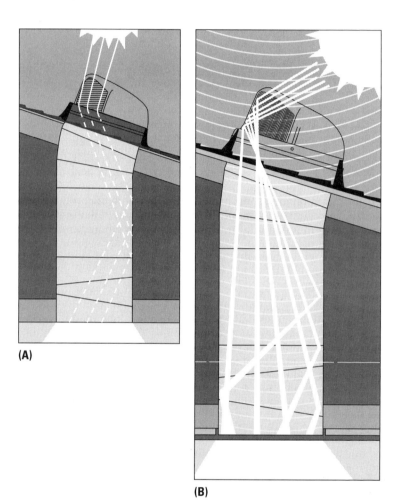

(A)

(B)

Figure 12-11   (A) The dome above the roof line; (B) the dome reflects the sunlight coming from any angle throughout the day in any season. *(Courtesy of ODL, Inc.)*

9. Install the ceiling trim ring.

10. Attach the diffuser. In the attic, assemble, adjust, and install the tubular components. In colder climates, it is necessary to insulate the tube shaft.

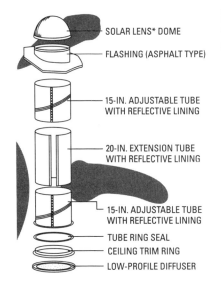

SOLAR LENS* DOME

FLASHING (ASPHALT TYPE)

15-IN. ADJUSTABLE TUBE
WITH REFLECTIVE LINING

20-IN. EXTENSION TUBE
WITH REFLECTIVE LINING

15-IN. ADJUSTABLE TUBE
WITH REFLECTIVE LINING

TUBE RING SEAL
CEILING TRIM RING
LOW-PROFILE DIFFUSER

Figure 12-12   Exploded view of the skylight. *(Courtesy of ODL, Inc.)*

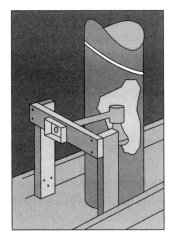

Figure 12-13   Conversion of skylight to a light fixture. *(Courtesy of ODL, Inc.)*

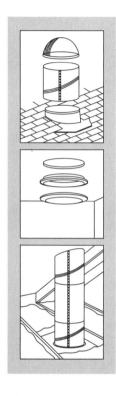

Figure 12-14 Installation of the skylight.
(Courtesy of ODL, Inc.)

Summary

Skylights are becoming more popular in residential dwellings. Skylights come double insulated, and some can be opened for ventilation.

A skylight is often placed at the top of a flight of stairs leading to the roof, or in an inside room where no side windows are available. Skylights may also be used in areas where the sun penetrates the glass for heating rooms. In some cases, the sun is used to heat water stored in ceiling tanks.

Review Questions

1. What are some of the advantages of a skylight?
2. Why are skylights used in some dwellings?
3. Where are skylights placed in dwellings?
4. What are some disadvantages of skylights?

5. Where are skylights located?

6. What makes it easy to install tube-type skylights?

7. How often should you check the skylight after installation?

8. How do you clean the mastic off the dome and other parts of the skylight?

9. How do you protect the plastic dome on the skylight?

10. Why would you want a skylight that opens to the outside?

Chapter 13

Cornice Details

The *cornice* is that projection of the roof at the eaves that forms a connection between the roof and the sidewalls. Following are four general types of cornice construction:

* Box
* Closed
* Wide box
* Open

Box Cornices

The typical *box cornice* (see Figure 13-1) utilizes the rafter projection for nailing surfaces for the facia and soffit boards. The soffit provides a desirable area for inlet ventilators. A frieze board is often used at the wall to receive the siding. In climates where snow and ice dams may occur on overhanging eaves, the soffit of the cornice may be sloped outward and left open $1/4$ inch at the facia board for drainage.

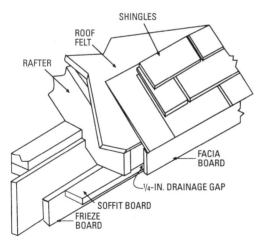

Figure 13-1   Box-cornice construction.

Closed Cornices

The *closed cornice* (see Figure 13-2) has no rafter projection. The overhang consists only of a frieze board and a shingle or crown

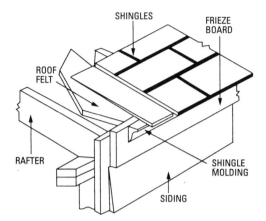

SHINGLES

FRIEZE
BOARD

ROOF
FELT

RAFTER

SHINGLE
MOLDING

SIDING

Figure 13-2   Closed-cornice construction.

molding. This type is not so desirable as a cornice with a projection, because it gives less protection to the sidewalls.

Wide Box Cornices

The *wide box cornice* (see Figure 13-3) requires forming members called *lookouts*, which serve as nailing surfaces and supports for the soffit board. The lookouts are nailed at the rafter ends and are toe-nailed to the wall sheathing and directly to the studs. The soffit can be of various materials (such as beaded ceiling, plywood, or bevel siding). A bed molding may be used at the juncture of the soffit and frieze. This type of cornice is often used in hip-roofed houses, and the facia board usually carries around the entire perimeter of the house.

Open Cornices

The *open cornice* (see Figure 13-4) may consist of a facia board nailed to the rafter ends. The frieze is either notched or cut out to fit between the rafters and is then nailed to the wall. The open cornice is often used for a garage. When it is used on a house, the roof boards are visible from below from the rafter ends to the wall line, and should consist of finished material. Dressed or matched V-beaded boards are often used.

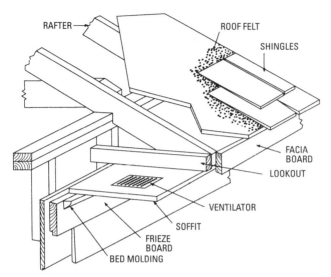

Figure 13-3   Wide cornice construction.

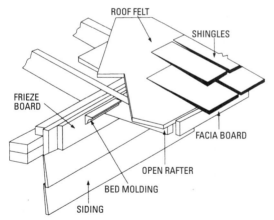

Figure 13-4   Open cornice construction.

Cornice Returns

The *cornice return* is the end finish of the cornice on a gable roof. The design of the cornice return depends to a large degree on the rake or gable projection, and on the type of cornice used. In a closed rake (a gable end with very little projection), it is necessary to use a frieze or rake board as a finish for siding ends (see Figure 13-5). This board is usually 1 1/8 inches thick and follows the roof slope to meet the return of the cornice facia. Crown molding or other type of finish is used at the edge of the shingles.

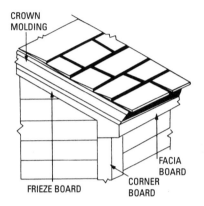

CROWN
MOLDING

FACIA
BOARD

FRIEZE BOARD

CORNER
BOARD

Figure 13-5 Closed cornice return.

When the gable end and the cornice have some projection (see Figure 13-6), a box return may be used. Trim on the rake projection is finished at the cornice return. A wide cornice with a small gable

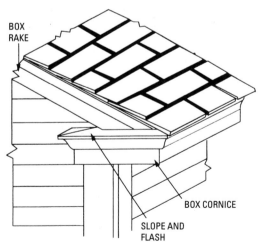

BOX
RAKE

BOX CORNICE

SLOPE AND
FLASH

Figure 13-6 Box cornice return.

projection may be finished as shown in Figure 13-7. Many variations of this trim detail are possible. For example, the frieze board at the gable end might be carried to the rake line and mitered with a facia board of the cornice. This siding is then carried across the cornice end to form a return.

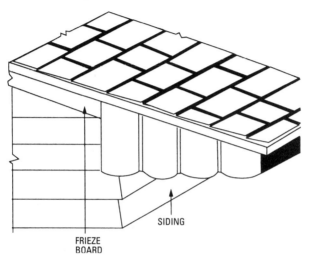

SIDING

FRIEZE
BOARD

Figure 13-7 Wide cornice return.

Rake or Gable-End Finish

The rake section is that trim used along the gable end of a house. Following are three general types commonly used:

- Closed
- Box with a projection
- Open

The *closed rake* (see Figure 13-8) often consists of a frieze or a rake board with a crown molding as the finish. A 1-inch × 2-inch square edge molding is sometimes used instead of the crown molding. When fiberboard sheathing is used, it is necessary to use a narrow frieze board that will leave a surface for nailing the siding into the end rafters.

If a wide frieze is used, nailing blocks must be provided between the studs. Wood sheathing does not require nailing blocks. The trim used for a box-rake section requires the support of the projected roof

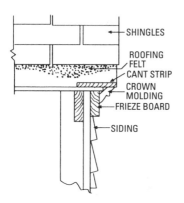

Figure 13-8 Closed-end finish at the rake.

boards (see Figure 13-9). In addition, lookouts or nailing blocks are fastened to the sidewall and to the roof sheathing. These lookouts serve as a nailing surface for both the soffit and the facia boards. The ends of the roof boards are nailed to the facia. The frieze board is nailed to the sidewall studs, and the crown and bed moldings complete the trim. The underside of the roof sheathing of the open-projected rake (see Figure 13-10) is generally covered with liner-boards (such as $5/8$-inch beaded ceiling). The fascia is held in place by nails through the roof sheathing.

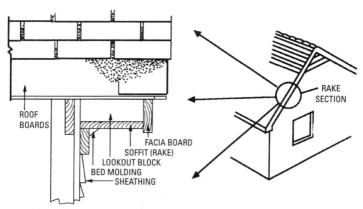

Figure 13-9 Box-end finish at the rake.

Summary

The cornice is that part of the roof at the eaves that forms a connection between the roof and sidewalls. There are generally four styles of cornice construction: box, closed, wide box, and open.

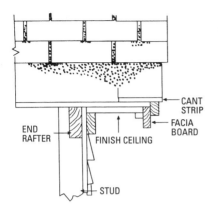

Figure 13-10 Open-end finish at the rake.

The box cornice construction generally uses the rafter ends as a nailing surface for the facia and soffit board. A board called the frieze board is used at the wall to start the wood siding. Wide box cornices require framework called lookouts, which serve as nailing surfaces and support the soffit board. The lookouts are nailed at the rafter end and nailed at the other end to the wall stud.

On the closed cornice, there is no rafter projection. There is no overhang, only a frieze board and molding. There is no protection from the weather for the sidewalls with this type of construction.

Review Questions

1. Name four types of cornice construction.
2. What is a frieze board?
3. Explain the purpose of the facia board.
4. What is the lookout block and when is it used?
5. What is the soffit board?
6. The _____ cornice has no rafter projection.
7. The _____ cornice requires forming members called lookouts.
8. The _____ cornice may consist of a facia board nailed to the rafter ends.
9. True or false, the cornice return is the end finish of the cornice on a gable roof.
10. The _____ section is that trim used along the gable end of a house.

Chapter 14

Doors

Doors (both exterior and interior) may be considered sash, flush, or louver. Flush doors may also be solid core or hollow core.

Manufactured Doors

For all practical purposes, doors can be obtained from the mill in stock sizes much cheaper than they can be made by hand. Stock sizes of doors cover a wide range, but those most commonly used are 2 feet, 4 inches × 6 feet, 8 inches; 2 feet, 8 inches × 6 feet, 8 inches; 3 feet, 0 inches × 6 feet, 8 inches and 3 feet, 0 inches × 7 feet, 0 inches. These sizes are either $1^3/_8$-inches (interior) or $1^3/_4$-inches (exterior) thick.

Sash and Paneled Doors

Panel and sash doors have for component parts a top rail, bottom rail, and two stiles that form the sides of the door (see Figure 14-1).

The rails and stiles of a door are generally mortised-and-tenoned, the mortise being cut in the side stiles (see Figure 14-2). Top and bottom rails on paneled doors differ in width, with the bottom rail being considerably wider. Intermediate rails are usually the same width as the top rail. Paneling material is usually plywood (which is set in grooves or dadoes in the stiles and rails), with the molding attached on most doors as a finish.

Flush Doors

Flush doors are usually perfectly flat on both sides. Solid planks are rarely used for flush doors. Flush doors are made with solid or hollow cores with two or more plies of veneer glued to the cores.

Solid-Core Doors

Solid-core doors are made of short pieces of wood glued together with the ends staggered very much like in brick laying. One or two plies of veneer are glued to the core. The first section (about $1/_8$ inch thick) is applied at right angles to the direction of the core, and the other section, $1/_8$ inch or less, is glued with the grain vertical. A $3/_4$-inch strip (the thickness of the door) is glued to the edges of the door on all four sides. Figure 14-3 shows this type of door construction.

Hollow-Core Doors

Hollow-core doors (which are flush) have wooden grids or other honeycomb material for the base, with solidwood edging strips on

Figure 14-1   Sash door with glazed sash.

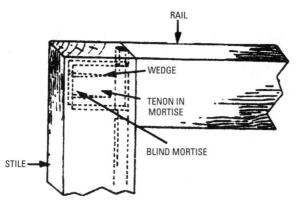

Figure 14-2   Door constructions showing mortise joints.

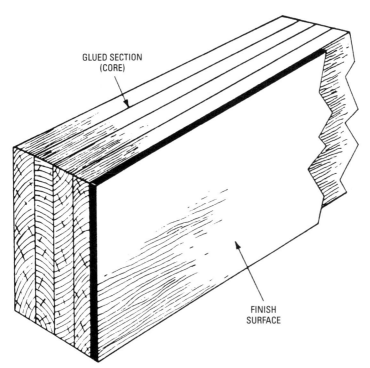

GLUED SECTION
(CORE)

FINISH
SURFACE

Figure 14-3   Construction of a laminated or veneered door.

all four sides. The face of this type door is usually 3-ply veneer instead of two single plies. The hollow-core door has a solid block on both sides for installing doorknobs and to permit the mortising of locks. The honeycomb-core door is for *interior* use only.

Louver Doors
This type of door has either stationary or adjustable louvers. It may be used as an interior door, room divider, or closet door. The louver door comes in many styles, such as those shown in Figure 14-4.

Installing Mill-Built Doors
A door frame may be constructed in numerous ways. A door frame consists of the following essential parts (see Figure 14-5):

- Sill
- Threshold

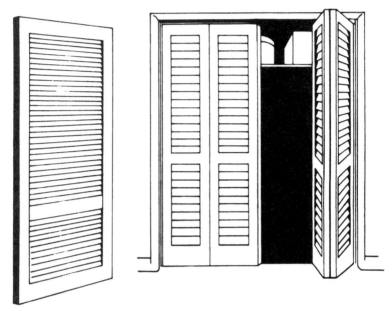

Figure 14-4   Two styles of louver doors.

- Side and top jamb
- Casing

Door Frames

Before the exterior siding is placed on the outside walls, the door openings are prepared for the frames. To prepare the openings, square off any uneven pieces of sheathing and wrap heavy building paper around the sides and top. Since the sill must be worked into a portion of the subflooring, no paper is put on the floor. Position the paper from a point even with the inside portion of the stud to a point about 6 inches on the sheathed walls, and staple it down.

Outside door frames are constructed in several ways. In more-hasty constructions, there will be no door frame. The studs on each side of the opening act as the frame and the outside casing is applied to the walls before the door is hung. The inside door frame is constructed the same way as the outside frame.

Doorjambs

Doorjambs are the lining to the framing of a door opening. Casings and stops are nailed to the jamb, and the door is securely fastened

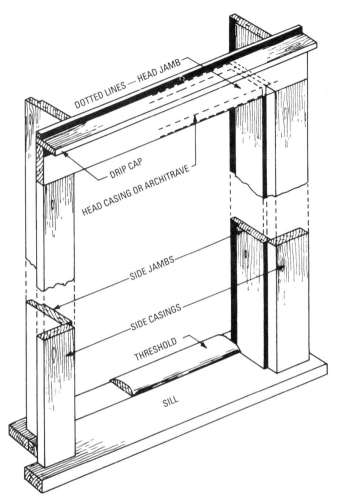

DOTTED LINES — HEAD JAMB

DRIP CAP

HEAD CASING OR ARCHITRAVE

SIDE JAMBS

SIDE CASINGS

THRESHOLD

SILL

Figure 14-5   View of a door frame showing the general construction.

by hinges at one side. The width of the jamb will vary in accordance with the thickness of the walls. Doorjambs are made and set in the following manner:

1. Regardless of how carefully the rough openings are made, be sure to plumb the jambs and level the heads when the jambs are set.

2. Rough openings are usually made $2^1/_2$ inches larger each way than the size of the door to be hung. For example, a 2-foot, 8-inch × 6-foot, 8-inch door would need a rough opening of 2 feet, $10^1/_2$ inches × 6 feet, $10^1/_2$ inches. This extra space allows for the jamb, the wedging, and the clearance space for the door to swing.

3. Level the floor across the opening to determine any variation in floor heights at the point where the jamb rests on the floor.

4. Cut the head jamb with both ends square, allowing for the width of the door plus the depth of both dadoes and a full $3/_{16}$ inch for door clearance.

5. From the lower edge of the dado, measure a distance equal to the height of the door plus the clearance wanted at the bottom.

6. Do the same thing on the opposite jamb. Only make additions or subtractions for the variation in the floor.

7. Nail the jambs and jamb heads together through the dado into the head jamb (see Figure 14-6).

8. Set the jambs into the opening and place small blocks under each jamb on the subfloor just as thick as the finish floor will be. This will allow the finish floor to go under the door.

9. Plumb the jambs and level the jamb head.

10. Wedge the sides to the plumb line with shingles between the jambs and the studs, and then nail securely in place.

11. Take care not to wedge the jambs unevenly.

12. Use a straightedge 5 to 6 feet long inside the jambs to help prevent uneven wedging.

13. Check each jamb and the head carefully. If a jamb is not plumb, it will have a tendency to swing the door open or shut, depending on the direction in which the jamb is out of plumb.

Door Trim

Door-trim material is nailed onto the jambs to provide a finish between the jambs and the wall material. This is called the *casing*. Sizes vary from $1/_2$ to $3/_4$ inch in thickness, and from $2^1/_2$ to 6 inches in width. Most casing material has a concave back, to fit over uneven wall material. In miter work, care must be taken to make all joints clean, square, neat, and well-fitted. If the trim is to be mitered at the top corners, a miter box, miter square, hammer, nail set, and block plane will be needed. Door openings are cased up in the following manner:

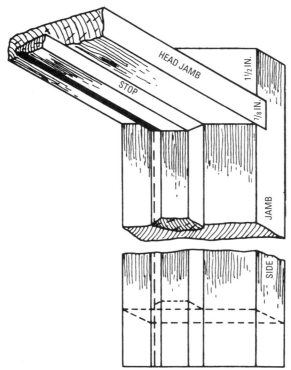

Figure 14-6 **Details showing upper head-jamb dadoes into side jamb.**

1. Leave a $1/4$-inch margin between the edge of the jamb and the casing on all sides.
2. Cut one of the side casings square and even with the bottom of the jamb.
3. Cut the top or mitered end next, allowing $1/4$-inch extra length for the margin at the top.
4. Nail the casing onto the jamb and set it even with the $1/4$-inch margin line, starting at the top and working toward the bottom.
5. Nails along the outer edge will need to be long enough to penetrate the casing and wall stud.
6. Set all nail heads about $1/8$-inch below the surface of the wood.

7. Apply the casing for the other side of the door opening in the same manner, followed by the head (or top) casing.

Hanging Doors

If flush or sash doors are used, install them in the finished door opening as described here:

1. Cut off the stile extension (if any) and place the door in the frame. Plane the edges of the stiles until the door fits tightly against the hinge side and clears the lock side of the jamb by about $1/16$ inch. Be sure that the top of the door fits squarely into the rabbeted recess and that the bottom swings free of the finished floor by about $1/2$ inch. The lock stile of the door must be beveled slightly so that the edge of the door will not strike the edge of the doorjamb.

2. After the proper clearance of the door has been made, set the door in position and place wedges as shown in Figure 14-7. Mark the position of the hinges on the stile and on the jamb with a sharp pointed knife. The lower hinge must be placed slightly above the lower rail of the door. The upper hinge of the door must be placed slightly below the top rail to avoid cutting out a portion of the tenons of the door rails. There are three measurements to mark: the location of the hinge on the jamb, the location of the hinge on the door, and the thickness of the hinge on both the jamb and the door.

3. Mortise the door butt hinges into the door and frame (see Figure 14-8). Three hinges are usually used on full-length doors to prevent warping and sagging.

4. Use the butt as a pattern, mark the dimension of the butts on the door edge and the face of the jamb. The butts must fit snugly and exactly flush with the edge of the door and the face of the jamb. A device called a *butt marker* can be helpful here.

After placing the hinges and hanging the door, mark off the position for the lock and handle. The lock is generally placed about 36 inches from the floor level. Hold the lock in position on the stile and mark off with a sharp knife the area to be removed from the edge of the stile. Mark off the position of the doorknob hub. Bore out the wood to house the lock, and chisel the mortises clean. After the lock assembly has been installed, close the door and mark the jamb for the striker plate.

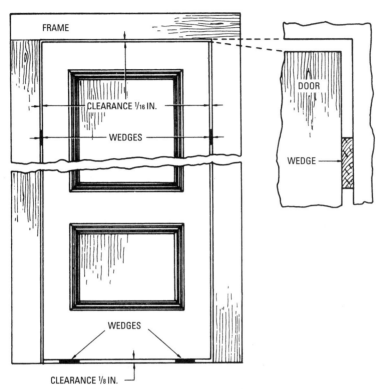

Figure 14-7 Sizing a door for an opening.

Swinging Doors

Frequently, it is desirable to hang a door so that it opens as you pass through from either direction, yet remains closed at all other times. For this purpose, you can use swivel-style spring hinges. This type of hinge attaches to the rail of the door and to the jamb like an ordinary butt hinge. Another type is mortised into the bottom rail of the door and is fastened to the floor with a floor plate. In most cases, the floor-plate hinge (see Figure 14-9) is best because it will not weaken and let the door sag. It is also designed with a stop to hold the door open at right angles, if so desired.

Sliding Doors

Sliding doors are usually used for walk-in closets. They take up very little space, and they allow a wide variation in floor plans. This type

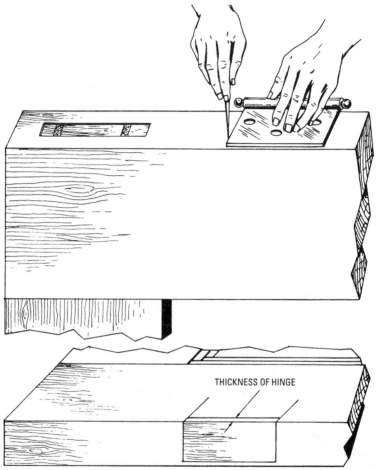

THICKNESS OF HINGE

Figure 14-8 Marking for hinges.

of door usually limits the access to a room or closet unless the doors are pushed back into a wall. Very few sliding doors are pushed back into the wall because of the space and expense involved. Figure 14-10 shows a double and a single sliding door track.

Garage Doors

Garage doors are made in a variety of sizes and designs. The principal advantage of any garage door is, of course, that it can be rolled

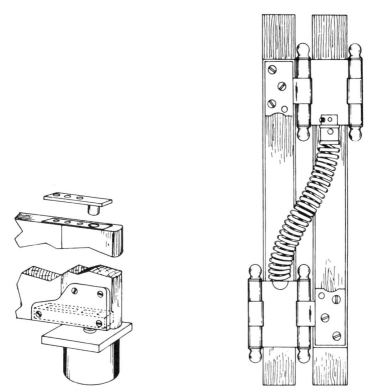

Figure 14-9   Two kinds of swivel-type spring hinges.

up out of the way. In addition, the door cannot be blown shut by the wind, and it is not obstructed by snow and ice.

Standard residential garage doors are usually 9 feet × 7 feet for a single-car garage and 16 feet × 7 feet for a double. Residential-type garage doors are usually $1^3/_4$-inches thick.

When ordering doors for the garage, the following information should be forwarded to the manufacturer:

- Width of opening between the finished jambs
- Height of the ceiling from the finished floor under the door to the underside of the finished header

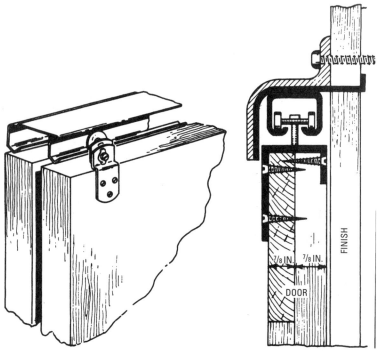

Figure 14-10 Two types of sliding-door tracks.

Figure 14-11 Typical 18-foot overhead residential garage door.

(A) Fiberglass.

(B) Steel.

(C) Wood.

Figure 14-12   Three types of garage doors.

- Thickness of the door
- Design of the door (number of glass windows and sections)
- Material of jambs (they must be flush)
- Headroom from the underside of the header to the ceiling, or to any pipes, lights, and so forth
- Distance between the sill and the floor level
- Proposed method of anchoring the horizontal track
- Depth to the rear from inside of the upper jamb
- Inside face width of the jamb buck, angle, or channel

This information applies for overhead doors only. It does not apply to garage doors of the sliding, folding, or hinged type. Doors can be furnished to match any style of architecture and may be provided with suitable size windows if desired (see Figure 14-11).

If your garage is attached to your house, the door often represents from one-third to one-fourth of the face of your house. Style and material should be considered to accomplish a pleasant effect with masonry or wood architecture. Figure 14-12 shows three types of overhead garage doors that can be used with virtually any kind of architectural design. Many variations can be created from combinations of raised panels with routed or carved designs (see Figure 14-13). These panels may also be combined with plain raised panels to provide other dramatic patterns and color combinations.

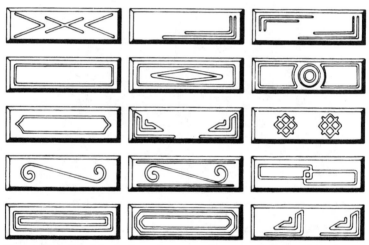

Figure 14-13   Variations in carved or routed panel designs.

Automatic garage-door openers were once a luxury item. However, recently the price has been reduced and failure minimized to the extent that many new installations include this feature. Automatic garage-door openers save time and eliminate the need to stop the car and get out in all kinds of weather. You also save the energy and effort required to open and close the door by hand.

The automatic door opener is a radio-activated, motor-driven power unit that mounts on the ceiling of the garage and attaches to the inside top of the garage door (see Figure 14-14). Electric impulses from a wall-mounted pushbutton, or radio waves from a transmitter in your car, start the door mechanism. When the door reaches its limit of travel (up or down), the unit turns off and awaits the next

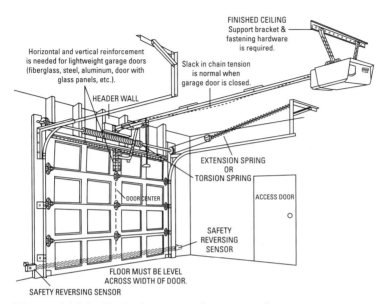

Figure 14-14 Typical automatic garage-door opener. *(Courtesy of Stanley Door Corporation)*

command. Most openers on the market have a safety factor built in. If the door encounters an obstruction in its travel, it will instantly stop, or stop and reverse its travel. The door will not close until the obstruction has been removed. When the door is completely closed, it is automatically locked and cannot be opened from the outside, making it burglar resistant. The unit has a light, which turns on when the door opens to light up the inside of the garage.

Summary

Most doors (both exterior and interior) are classified as sash, flush, or louver.

Sash and panel doors are made in many styles. The rails and stiles are generally mortised-and-tenoned. Top and bottom rails on paneled doors differ in width, with the bottom rail considerably wider. The center rail is generally the same width as the top rail. The panel material is usually plywood, which is set in grooves or dadoes in the stiles and rails.

Solid-core doors are made of short pieces of wood glued together with the ends staggered very much like brick laying. Hollow-core doors have wooden grids or some type of honeycomb material for the base, with solid wood edging strips on all four sides. Glued to the cores of these doors are two or three layers of wood veneer, which make up the door panel. The honeycomb-core door is made for interior use only.

Review Questions

1. Name the various types of doors.

2. Why are honeycomb-core doors made for interior use only?

3. What is a doorstop?

4. When hanging a door, how much clearance should there be at top, bottom, and sides?

5. How are solid-core doors constructed?

6. The rails and stiles of a door are generally _____ -and-tenoned.

7. The doorframe consists of the following essential parts: sill, threshold, side and top jambs, and _____.

8. Doorjambs are the _____ to the framing of a door opening.

9. What is the door casing?

10. True or false: most casing material has a convex back.

Chapter 15

Windows

Windows in any building structure not only provide a means for illuminating the interior, but also provide a decorative touch to the structure. Following are the three *main* window types:

- Double-hung
- Casement
- Gliding

There are also *awning*, *bow*, and *bay* windows (see Figure 15-1). Windows consist essentially of two parts:

- Frame
- Sash

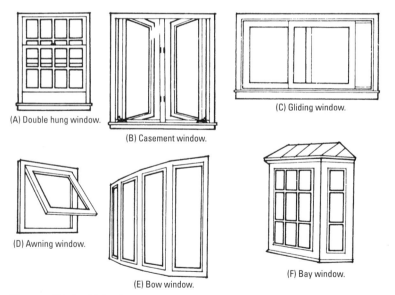

(A) Double hung window.

(B) Casement window.

(C) Gliding window.

(D) Awning window.

(E) Bow window.

(F) Bay window.

Figure 15-1   Window types.

The frame is made up of four basic parts: the *head*, two *jambs*, and the *sill*.

Window Framing

The window *sash* fits into the window frame. It is set into a rough opening in the wall framing and is intended to hold the sash in place.

Double-Hung Windows

The double-hung window holds two pieces of sash: (an upper and lower), which slide vertically past each other (see Figure 15-2).

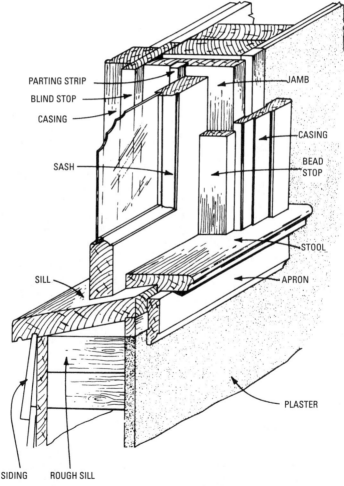

Figure 15-2   Three-quarter view of window frame.

This type of window has some advantages and some disadvantages. Screens can be installed on the outside of the window without interfering with its operation. For full ventilation of a room, only half of the area of the window can be utilized, and any current of air passing across its face is, to some extent, lost in the room. Double-hung windows are sometimes more involved in their frame construction and operation than the casement window. Ventilation fans and air conditioners can be placed in the window with it partly closed.

Hinged or Casement Windows

There are two types of casement windows:

- Out swinging
- In swinging

These windows may be hinged at the side, top, or bottom. The casement window that opens out requires the screen to be located on the inside. This type of window, when closed, is more efficient as far as waterproofing. The in-swinging casement windows, like double-hung windows, are clear of screens, but they are extremely difficult to make watertight. Casement windows have the advantage of their entire area being opened to air currents, thus catching a parallel breeze and deflecting it into a room. Casement windows are considerably less complicated in their construction than double-hung units are. Sill construction is very much like that for a double-hung window, however, but with the stool much wider and forming a stop for the bottom rail of the sash. When there are two casement windows in a row in one frame, they are separated by a vertical double jamb (called a *mullion*), or the stiles may come together in pairs like a French door. The edges of the stiles may be a reverse rabbet, a beveled reverse rabbet with battens, or beveled astrogals. The battens and astrogals ensure better weather tightness.

Gliding, Bow, Bay, and Awning Windows

Gliding windows consist of two sashes that slide horizontally right or left. They are often installed high up in a home to provide light and ventilation without sacrificing privacy.

Awning windows have a single sash hinged at the top and open outward from the bottom. They are often used at the bottom of a fixed picture window to provide ventilation without obstructing the view. They are popular in ranch homes.

Bow and bay windows add architectural interest to a home. Bow windows curve gracefully, while bay windows are straight across the middle and angled at the ends. They are particularly popular in Georgian-style and Colonial-style homes.

Wood windows are better than metal ones for insulation purposes, simply because metal conducts heat better than wood. Nevertheless, even more important is double-glazing, which contains a dead air space that inhibits heat escaping (or getting in, should you have air conditioning). The second pane can be incorporated in the window (see Figure 15-3) or it can be removable. If you live in an area where heating costs are very high, consider triple glazing (three panes of glass with air spaces between). Tinted or reflective glass is good for warding off the sun's rays in warmer climates.

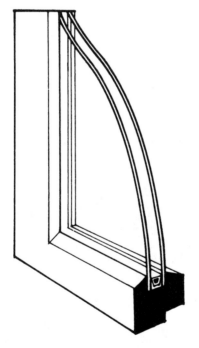

Figure 15-3   Insulated glass.

Window Sash

A sash is a framework that holds the glass lights. The lights are divided by thin strips called *muntins*. There are two general types of wood sash: fixed (or permanent) and movable. Fixed window sash are removable only with the aid of a carpenter. Movable sash may be of the variety that slide up and down in channels in the frame (called *double-hung*). Casement-type sash swing in or out and are hinged on the sides.

Sash Installation

Place the upper double-hung sash in position and trim off a slight portion of the top rail to ensure a good fit, and tack the upper sash in position. Fit the lower sash in position by trimming off the sides. Place the lower sash in position, and trim off a sufficient amount from the bottom rail to permit the meeting rails to meet on a level. In most cases, the bottom rail will be trimmed on an angle to permit the rail and sill to match inside and outside (see Figure 15-4).

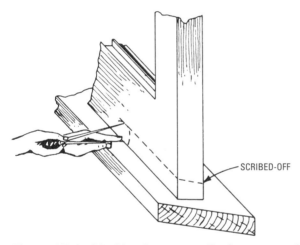

SCRIBED-OFF

Figure 15-4   Marking bottom-rail trim to match sill plate.

Sash Weights

If sash weights are used, remove each sash after it has been properly cut and sized. Select sash weights equal to one-half the weight of each sash and place in position in the weight pockets. Measure the proper length of sash cord for the lower sash and attach it to the stiles and weights on both sides. Adjust the length of the cord so that the weight will not strike the pulley or bottom of the frame when the window is moved up and down. Install the cords and weights for the upper sash and adjust the cord so that the weights run smoothly. Close the pockets in the frame and install the blind stop, parting strip, and bead stop.

There are many other types of window lifts (such as spring-loaded steel tapes, spring-tension metal guides, and full-length coil springs).

Glazing Sash

The panes of glass (or *lights*, as they are called) are generally cut $1/8$ inch smaller on all four sides to allow for irregularities in cutting and in the sash. This leaves an approximate margin of $1/16$ inch between the edge of the glass and the sides of the rabbet. Figure 15-5 shows two lights or panes of glass in position for glazing. To install the window glass properly, first spread a film of glazing compound close to the edge on the inside portion of the glass. After the glass has been inserted, drive or press in at least two glaziers' points on each side (see Figure 15-6).

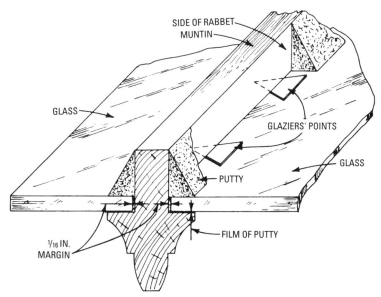

Figure 15-5   Glaziers' points, which are removed to replace broken glass.

When the glass is firmly secured with the glaziers' points, the compound (which is soft) is put on around the glass with a putty knife and beveled (see Figure 15-6). Do not project the compound beyond the edge of the rabbet so that it will be visible from the other side.

Figure 15-7 shows double-hung windows, with small panes simulated by having spacers inserted between the two pieces of glass that make up the window. This way, the windows are easier to clean, with one piece of glass rather than numerous small panes.

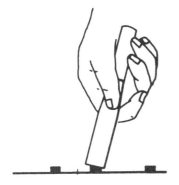

Figure 15-6 Push glaziers' points in with putty knife.

Figure 15-7 Double-hung windows capped off with semicircular stationary "half-moons."

Note the stationary-type semicircle windows on top to add style to the double-hung box appearance. Windows of all shapes and sizes are now available, so architects have great freedom in their design work.

Shutters

In coastal areas where damaging high winds occur frequently, shutters are necessary to protect large plate-glass windows from being broken. The shutters are mounted on hinges and can be closed at a moment's notice. Throughout the Midwest, shutters are generally installed for decoration only and are mounted stationary to the outside wall. There are generally two types of shutters: the solid panel and the slat (or louver) type. Louver shutters can have stationary or movable slats.

Summary

Many styles and sizes of windows are used in various house designs, but the main ones are double-hung, casement, and gliding. A window consists generally of two parts: the frame and the sash.

Double-hung windows are made up of three parts: the upper and lower sash (which slide vertically past each other) and the frame. Only half of the area of the window can be used for ventilation, which is a disadvantage.

Shutters serve a purpose near the coastline, but are only decorative in most of the country.

Review Questions

1. Name the various window classifications.
2. What size should the rough opening be for a double-hung window?
3. What are some advantages of casement windows?
4. Name a few advantages in using window shutters.
5. What are glazier points, and why should they be used when installing window glass?
6. Gliding windows consist of two sash that slide _____ right and left.
7. A _____ is a framework that holds the glass lights.
8. If sash _____ are used, remove each sash after it has been properly cut and sized.
9. How do you install glazers' points?
10. True or false: the panes of glass, or strips, as they are called, are generally cut $1/8$ inch smaller on all four sides.

Chapter 16

Siding

Sheathing is nailed directly to the framework of the building. Its purpose is to strengthen the building, to provide a base material to which finish siding can be attached, to act as insulation, and, in some cases, to be a base for further insulation. Some of the common types of sheathing include fiberboard, wood, and plywood.

Fiberboard Sheathing

Fiberboard usually comes in 2-foot × 8-foot or 4-foot × 8-foot sheets that are tongue-and-grooved and generally coated or impregnated with an asphalt material that increases water resistance. Thickness is normally $^1/_2$ or $^{25}/_{32}$ inch, and fiberboard may be used where the stud spacing does not exceed 16 inches. Fiberboard sheathing should be nailed with 2-inch galvanized roofing nails or other type of noncorrosive nails. If the fiberboard is used as sheathing, most builders will use plywood at all corners, in the same thickness as the sheathing, to strengthen the walls (see Figure 16-1).

Wood Sheathing

Wood wall sheathing can be obtained in almost all widths, lengths, and grades. Generally, widths are from 6 to 12 inches, with lengths selected for economical use. Almost all solid-wood wall sheathing used is $^{25}/_{32}$ to 1 inch in thickness. This material may be nailed on horizontally or diagonally (see Figure 16-2). Wood sheathing is laid on tight, with all joints made over the studs. If the sheathing is to be put on horizontally, it should be started at the foundation and worked toward the top. If the sheathing is installed diagonally, it should be started at the corners of the building and worked toward the center or middle.

Diagonal sheathing should be applied at a 45° angle. This method of sheathing adds greatly to the rigidity of the wall and eliminates the need for corner bracing. It also provides an excellent tie to the sill plate when it is installed diagonally. There is more lumber waste than with horizontal sheathing because of the angle cut, and the application is somewhat more difficult. Figure 16-3 shows the wrong way and the correct way of laying diagonal sheathing.

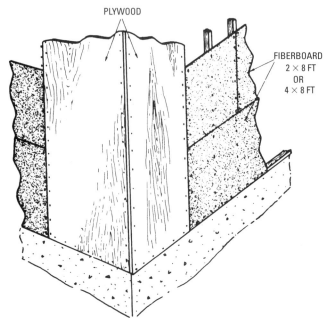

Figure 16-1 **Method of using plywood on all corners as bracing when using fiberboard as exterior sheathing.**

Plywood Sheathing

Plywood as a wall sheathing is highly recommended because of its size, weight, and stability, plus the ease and rapidity of installation (see Figure 16-4). It adds considerably more strength to the frame structure than the conventional horizontal or diagonal sheathing. When plywood sheathing is used, corner bracing can also be omitted. Large-size panels effect a major savings in the time required for application and still provide a tight, draft-free installation that contributes a high insulation value to the walls. The thickness of plywood wall sheathing is $1/2$ inch for 16-inch stud spacing, and $5/8$ inch to $3/4$ inch for 24-inch stud spacing. The panels should be installed with the face grain parallel to the studs. However, a little more stiffness can be obtained by installing them across the studs, but this requires more cutting and fitting. Nail spacing should not be more than 6 inches on the center at the edges of the panels, and not more than 12 inches on center elsewhere. Joints should meet on the centerline of framing members.

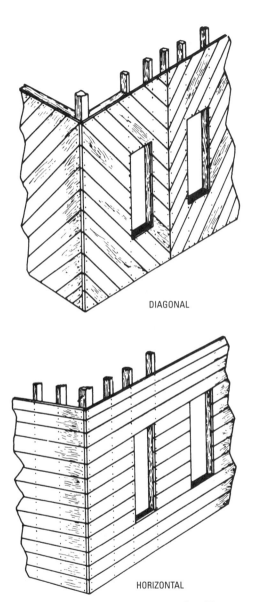

DIAGONAL

HORIZONTAL

Figure 16-2   Two methods of nailing on wood sheathing.

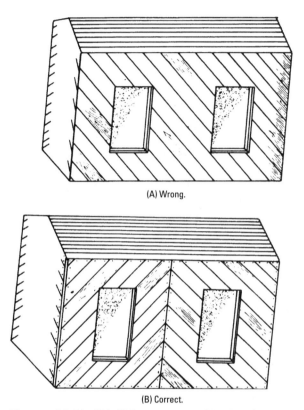

(A) Wrong.

(B) Correct.

Figure 16-3 (A) Wrong way of laying wood sheathing, and (B) correct way.

Urethane and Fiberglass

With the accent in recent years on saving energy, a number of other insulations have been developed that have high insulating value. For example, there is urethane, a $1^{1}/_{4}$ -inch-thick material that, when combined with regular insulation, yields a resistance (R) factor of 22 (see Figure 16-5). There is also fiberglass insulation with an R factor of 4.8. Such insulations are particularly good on masonry construction, because brick itself has very little insulating value and requires whatever insulation can be built in.

Sheathing Paper

Sheathing paper should be used on a frame structure when wood or plywood sheathing is used. It should be water-resistant, but

Figure 16-4   Plywood is a popular sheathing. *(Courtesy American Plywood Assn.)*

not vapor-resistant. These exterior air-infiltration barriers block air movement, but let water vapor pass. Two common brand names are Tyvek and Typar. It should be applied horizontally, starting at the bottom of the wall. Succeeding layers should lap about 4 inches, and lap over strips around openings. Strips about 6 inches wide should be installed behind all exterior trim or exterior openings.

Wood Siding

One of the materials most characteristic of the exteriors of American houses is wood siding. The essential properties required for wood siding are good painting characteristics, easy working qualities, and freedom from warp. These properties are present to a high degree in the cedars, Eastern white pine, sugar pine, Western white pine, cypress, and redwood.

Material used for exterior siding should preferably be of a select grade, and should be free from knots, pitch pockets, and wavy edges. Vertical-grain wood has better paint-holding and weathering characteristics than flat-grain wood. The moisture content at the

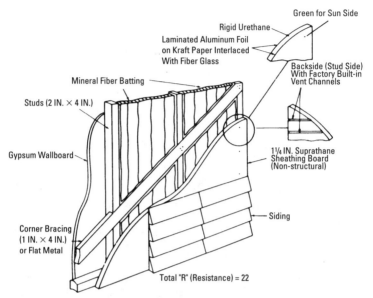

Figure 16-5 Urethane in combination with batt insulation here produces an R factor of 22.

time of application should be that which it would attain in service. This would be approximately 12 percent, except in the dry Southwestern states, where the moisture content should average about 9 percent.

Bevel Siding

Plain bevel siding (see Figure 16-6) is made in nominal 4-, 5-, and 6-inch widths with $^7/_{16}$-inch butts; and 6-, 8-, and 10-inch widths with $^9/_{16}$- and $^{11}/_{16}$-inch butts. Bevel siding is generally furnished in random lengths varying from 4 to 20 feet.

Drop siding is generally $^3/_4$ inch thick and is made in a variety of either patterns with matched or shiplap edges. Figure 16-7 shows three common patterns of drop siding that are applied horizontally. V-rustic siding (see Figure 16-7A) may be applied vertically (for example, at the gable ends of a house). Drop siding (see Figure 16-7B) is designed to be applied directly to the studs, and it thereby serves as sheathing and exterior-wall covering. It is widely used in this manner in farm structures (such as sheds and garages) in all parts of the country. When used over or in contact with other material

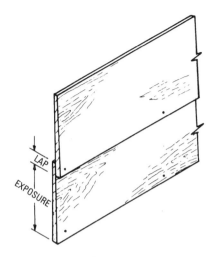

Figure 16-6   Bevel siding.

LAP

EXPOSURE

(such as sheathing or sheathing paper), water may work through the joints and be held between the sheathing and the siding. This sets up a condition conducive to paint failure and decay. Such problems can be avoided when the sidewalls are protected by a good roof overhang.

Square-Edge Siding

Square-edge, or clapboard, siding made of $^{25}/_{32}$-inch board is occasionally selected for architectural effects. In this case, wide boards are generally used. Some of this siding is also beveled on the back at the top to allow the boards to lie rather close to the sheathing, thus providing a solid nailing surface.

Vertical Siding

Vertical siding is commonly used on the gable ends of a house, over entrances, and sometimes for large wall areas. The type used may be plain-surfaced matched boards, patterned matched boards, or square-edge boards covered at the joint with a batten strip. Matched vertical siding should preferably not be more than 8 inches wide and should have 2 eight-penny nails not more than 4 feet apart. Backer blocks should be placed between studs to provide a good nailing base. The bottoms of the boards should be undercut to form a water drip.

Batten-type siding is often used with wide square-edged boards, which, because of their width, are subjected to considerable

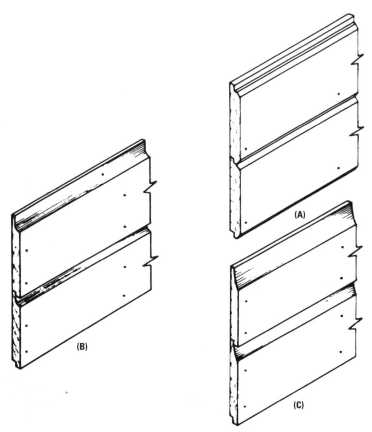

Figure 16-7 Types of drop siding: (A) V-rustic, (B) drop, (C) rustic drop.

expansion and contraction. The batten strips used to cover the joints should be nailed to only one siding board so the adjacent board can swell and shrink without splitting the boards or the batten strip.

Plywood Siding

Plywood is often used in gable ends, sometimes around windows and porches, and occasionally as an overall exterior wall covering. The sheets are made either plain or with irregularly cut striations. It can be applied horizontally or vertically. The joints can be molded

Table 16-1   Suggested Thickness of Plywood Siding

Minimum Thickness	Maximum Stud Space
$3/8$ inch	16 inches on center
$1/2$ inch	20 inches on center
$5/8$ inch	24 inches on center

batten, V-grooves, or flush. Sometimes it is installed as lap siding. Plywood siding should be of exterior grade. For unsheathed walls, the thickness shown in Table 16-1 is suggested.

Preservative Treatment

Houses are often built with little or no overhang of the roof, particularly on the gable ends. This permits rainwater to run down freely over the face of the siding. Under such conditions water may work up under the laps in bevel siding or through joints in drop siding by capillary action, providing a source of moisture that may cause paint blisters or peeling.

A generous application of a water-repellent preservative to the back of the siding will be quite effective in reducing capillary action with bevel siding. In drop siding, the treatment would be applied to the matching edges. Dipping the siding in the water repellent would be still more effective. The water repellent should be applied to all end cuts, at butt points, and where the siding meets door and window trim.

Wood Shingles and Shakes

Cedar shingles and shakes come in a variety of grades. They may be applied in several ways. You may get them in random widths 18 to 24 inches long, or in a uniform 18 inches. The shingles may be installed on regular sheathing or on an under course of shingles (which produces a shadowed effect). Cedar stands up to the weather well and does not have to be painted.

Installation of Siding

The spacing for siding should be carefully laid out before the first board is applied. The bottom of the board that passes over the top of the first-floor windows should coincide with the top of the window cap (see Figure 16-8). To determine the maximum board spacing or exposure, deduct the minimum lap from the overall width of the siding. The number of board spaces between the top of the

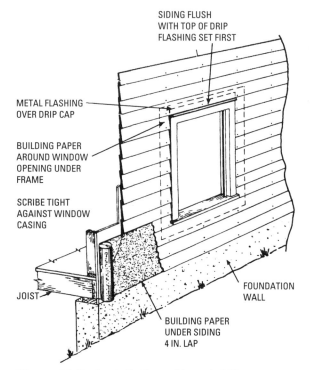

SIDING FLUSH
WITH TOP OF DRIP
FLASHING SET FIRST

METAL FLASHING
OVER DRIP CAP

BUILDING PAPER
AROUND WINDOW
OPENING UNDER
FRAME

SCRIBE TIGHT
AGAINST WINDOW
CASING

JOIST

FOUNDATION
WALL

BUILDING PAPER
UNDER SIDING
4 IN. LAP

Figure 16-8   Installation of bevel siding.

window and the bottom of the first course at the foundation wall should be such that the maximum exposure will not be exceeded. This may mean that the boards will have less than the maximum exposure.

Siding starts with the bottom course of boards at the foundation (see Figure 16-9). Sometimes the siding is started on a water table, which is a projecting member at the top of the foundation to throw off water (see Figure 16-10). Each succeeding course overlaps the upper edge of the lower course. The minimum head lap is 1 inch for 4- and 6-inch widths, and $1^{1}/_{4}$ inch for widths over 6 inches. The joints between boards in adjacent courses should be staggered as much as possible. Butt joints should always be made on a stud, or where boards butt against window and door casings and corner

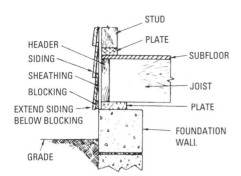

Figure 16-9 Installation of the first or bottom course.

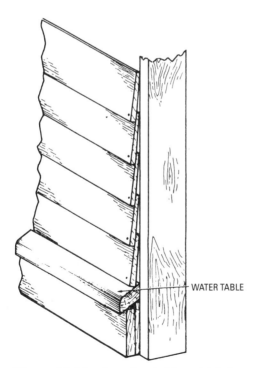

Figure 16-10 A water table, which is sometimes used.

boards. The siding should be carefully fitted and be in close contact with the member or adjacent pieces. Some carpenters fit the
boards so tight that they have to spring the boards in place, which
assures a tight joint. Loose-fitting joints allow water to get behind
the siding, thereby causing paint deterioration around the joints
and setting up conditions conducive to decay at the ends of the
siding.

Types of Nails

Nails cost very little compared to the cost of siding and labor, but
the use of good nails is important. It is poor economy to buy siding
that will last for years and then use nails that will rust badly within a
few years. Rust-resistant nails will hold the siding permanently and
will not disfigure light-colored paint surfaces.

There are two types of nails commonly used with siding, one
having a small head and the other a slightly larger head. The *small-
head casing nail* is set (driven with a nail set) about $1/16$ inch below the
surface of the siding. The hole is filled with putty after the prime coat
of paint is applied. The *large-head nail* is driven flush with the face
of the siding, with the head being later covered with paint. Ordinary
steel wire nails tend to rust in a short time and cause a disfiguring
stain on the face of the siding. In some cases, the small-head nail
will show rust spots through the putty and paint. Noncorrosive-
type nails (galvanized, aluminum, and stainless steel) that will not
cause rust stains are readily available.

Bevel siding should be face-nailed to each stud with noncorrosive
nails, the size depending upon the thickness of the siding and the
type of sheathing used. The nails are generally placed about $1/2$ inch
above the butt edge, in which case they pass through the upper edge
of the lower course of siding. Another method recommended for
bevel siding by most associations representing siding manufacturers
is to drive the nails through the siding just above the lap so that the
nail misses the thin edge of the piece of siding underneath. The latter
method permits expansion and contraction of the siding board with
seasonal changes in moisture content.

Corner Treatment

The method of finishing the wood siding at the exterior corners
is influenced somewhat by the overall house design. Corner boards
are appropriate to some designs, and mitered joints to others. Wood
siding is commonly joined at the exterior corners by corner boards,
mitered corners, or metal corners.

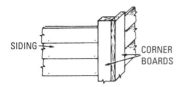

SIDING

CORNER
BOARDS

Figure 16-11 Corner treatment for bevel siding using the corner board.

Corner Boards

Corner boards (see Figure 16-11) are used with bevel or drop siding and are generally made of nominal 1- or 1¹/₄-inch material, depending upon the thickness of the siding. They may be either plain or molded, depending on the architectural treatment of the house. The corner boards may be applied vertically against the sheathing, with the siding fitting tightly against the narrow edge of the corner board. The joints between the siding and the corner boards and trim should be caulked or treated with a water repellent. Corner boards and trim around windows and doors are sometimes applied over the siding, a method that minimizes the entrance of water into the ends of the siding.

Mitered Corners

Mitered corners (see Figure 16-12) must fit tightly and smoothly for the full depth of the miter. To maintain a tight fit at the miter, it is important that the siding is properly seasoned before delivery, and is stored at the site to be protected from rain. The ends should be set in oil-based paint when the siding is applied, and the exposed

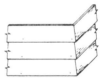

Figure 16-12 The mitered corner treatment.

faces should be primed immediately after it is applied. At interior corners, shown in Figure 16-13, the siding is butted against a corner strip of nominal 1- or 1¹/₄-inch material, depending upon the thickness of the siding.

Metal Corners

Metal corners (see Figure 16-14) are made of 8-gage metals (such as aluminum and galvanized iron). They are used with bevel siding as a substitute for mitered corners, and can be purchased at most lumberyards. The application of metal corners takes less skill than is required to make good mitered corners, or to fit the siding to a corner board. Metal corners should always be set in white lead paint.

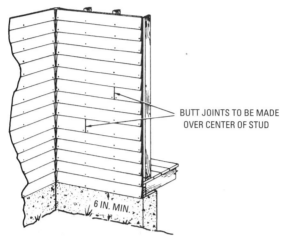

Figure 16-13 The construction of an interior corner using bevel siding.

Metal Siding

The metal most popular of those used in siding is aluminum. It is installed over most types of sheathing with an aluminum building paper (for insulation) nailed on between the sheathing and siding or insulation built onto the siding. Its most attractive characteristic is the long-lasting finish. The cost of painting and maintenance has made this type of siding doubly attractive.

Figure 16-14 Corner treatment for bevel siding using the corner metal caps.

Aluminum siding can be installed over old siding that has cracked and weathered, or where paint will not hold up.

Vinyl Siding

Also popular is vinyl siding (see Figure 16-15). This comes in a wide variety of colors, textures, and styles. As with aluminum siding, the big advantage of vinyl siding is that it does not need to be painted and will not corrode, dent, or pit. It is relatively susceptible to cracking if hit when it is very cold.

Figure 16-15 Solid vinyl siding comes in various colors and textures. It never needs to be painted.

Summary

Sheathing is nailed directly to the framework of the building. The purpose of sheathing is to strengthen the structure, to provide a nailing base for siding, and to act as insulation. Types of sheathing include fiberboard, wood, plywood, urethane, and fiberglass.

Fiberboard is generally furnished in 2-foot × 8-foot or 4-foot × 8-foot sheets and is usually coated with an asphalt material to make it waterproof. When fiberboard sheathing is used, most builders will use plywood at all corners to strengthen the walls. Fiberboard is normally $^1/_2$- or $^{25}/_{32}$-inch thick and generally tongue-and-grooved.

Wood sheathing is generally any size from 1 inch × 6 inches to 1 inch × 12 inches in width. The material may be installed horizontally or diagonally with all joints made over a stud. Diagonal sheathing should be applied at a 45° angle. This adds greatly to the rigidity of the walls and eliminates the need for corner bracing. More lumber waste is realized than when applying horizontal sheathing, but an excellent tie to the sill plate is accomplished when installed diagonally.

One of the most popular exterior-wall finishes of American houses is wood siding. Various types or styles include bevel, drop, square-edge, and vertical siding. A number of methods are used as a corner treatment when using wood bevel siding. Some corners are designed to use a vertical corner board, which is generally 1- or $1^1/_4$-inch material. Mitered corners are sometimes used, or the same effect can be obtained by using metal corners.

Of the metal sidings, aluminum is the most popular.

Review Questions

1. What is fiberboard and how is it used as sheathing?

2. What are some advantages in using wood sheathing placed diagonally?

3. Name the various styles of wood siding.

4. How are corners on wood siding treated?

5. What is a water table?

6. What is the advantage of having fiberboard with tongue-and-grooved edges?

7. At what angle is diagonal sheathing applied?

8. What is the advantage of applying sheathing diagonally?

9. Sheathing paper should be used on a frame structure when wood or _____ sheathing is used.

10. True or false: vertical siding is commonly used on the gable ends of a house.

Appendix A

Professional and Trade Associations

Table A-1 shows some professional and trade associations in the fields of doors and windows.

Table A-1   Professional and Trade Associations

Organization	Address	Web Site
American Architectural Manufacturers Association	1827 Walden Office Square, Suite 104, Schaumburg, IL 60175-4628	www.aamanet.org
American Hardware Manufacturers Association	801 North Plaza Drive, Schaumburg, IL 60175-4977	www.ahma.org
Builders Hardware Association	555 Lexington Avenue, 17th Floor, New York, NY 10017	www.builders hardware.com
Door and Hardware Institute	14150 Newbrook Drive, Suite 200, Chantilly, VA 20151-2225	www.dhi.org
Glass Association of North America	2945 SW Wanamaker Drive, Suite A, Topeka, KS 66614-5521	www.glasswebsite. com
National Fenestration Rating Council	1500 Spring Street, Suite 500, Silver Spring, MD 20910	www.nfrc.org
National Glass Association	8200 Greensboro Drive, McLean, VA 22102	www.glass.org
National Wood Window and Door Association	1400 East Touhy Avenue, Suite 470, Des Plaines, IL 60018	www.nwwda.org
Sealed Insulating Glass Manufacturing Association	401 North Michigan Avenue, Chicago, IL 60611	www.sigmaonline. org/sigma/
Steel Door Institute	50200 Detroit Road, Cleveland, OH 44145-1967	www.steeldoor.org
Steel Window Institute	1500 Sumner Avenue, Cleveland, OH 44115-2851	www.steelwindows. com

Index